BASIC MOLECULAR AND CELL BIOLOGY

Third edition

BASIC MOLECULAR AND CELL BIOLOGY

Third edition

BASIC MOLECULAR AND CELL BIOLOGY

Third edition

Edited by

David S Latchman

Professor of Molecular Pathology,
University College London Medical School, UK

BMJ
Publishing
Group

© BMJ Publishing Group 1988, 1993, 1997

First published in 1988
Reprinted 1988
Second edition 1993
Second impression 1994
Third impression 1995
Third edition 1997

British Library Cataloguing in Publication Data.
A catalogue record for this book is available from the British Library.

ISBN 0-7279-1195-3

Contents

List of contributors

John Armstrong
Senior Research Fellow, School of Biological Sciences, University of Sussex, Brighton

CR Bebbington
Head of Research, Oxford Biomedica, Oxford

Paul M Brickell
Professor of Molecular Haematology, Molecular Haematology Unit, Institute of Child Health, London

NH Carey
Intellectual Property Management, Chinnor, Oxfordshire

Kerry A Chester
Senior Lecturer in Oncology, CRC Targetting and Imaging Group, The Royal Free Hospital School of Medicine, London

Heung Chong
Clinical Research Fellow, Molecular Therapy Laboratory, Imperial Cancer Research Fund Molecular Oncology Unit, Hammersmith Hospital, London

BR Clark
Division of Clinical Sciences, Department of Medicine, University of Toronto, Canada

MKL Collins
Professor of Immunology, Department of Immunology, University College London Medical School, London

Ian NM Day
Lister Institute Research Fellow, Department of Medicine, University College London Medical School, The Rayne Institute, London

Professor TM Dexter FRS
Department of Experimental Haematology, University of Manchester, Paterson Institute of Cancer Research, Christie Hospital and Holt Radium Institute, Manchester

Professor DR Garrod
School of Biological Sciences, University of Manchester, Manchester

Robert E Hawkins
Professor of Oncology, Bristol Oncology Centre, Bristol

Meirion B Llewelyn
Senior Registrar in Infectious Diseases, Department of Medicine, University of Wales, Cardiff

RA Laskey FRS
Charles Darwin Professor of Animal Embryology, Wellcome/CRC Institute, University of Cambridge, Cambridge

David S Latchman
Professor of Molecular Biology, Medical Molecular Biology Unit, Department of Molecular Pathology, University College London Medical School, London

Durward Lawson
Reader in Cell Biology, Department of Molecular Pathology, University College London Medical School, London

Myra O McClure
Department of Genito-Urinary Medicine and Communicable Diseases, St Mary's Hospital Medical School, London

Professor AF Markham
West Riding Medical Research Trust, Molecular Medicine Unit, Department of Medicine, St James' University Hospital, University of Leeds, Leeds

Robert H Michell FRS
Royal Society Research Professor, Centre for Clinical Research in Immunology and Signalling, The Medical School, University of Birmingham, Birmingham

Paul Nurse FRS
Director of Research (Laboratories), Cell Cycle Laboratory, Imperial Cancer Research Fund, London

William Reardon
Senior Lecturer in Clinical Genetics and Dysmorphology, Mothercare Unit of Clinical Genetics and Fetal Medicine, Institute of Child Health and Great Ormond Street Hospital for Children, London

Stephen J Russell
MRC Senior Clinical Fellow, Cambridge Centre for Protein Engineering, Medical Research Council Centre, Cambridge

Richard G Vile
Head of Laboratory, Molecular Therapy Laboratory, Imperial Cancer Research Fund Molecular Oncology Unit, Hammersmith Hospital, London

Sir David J Weatherall FRS
Regius Professor of Medicine, Institute of Molecular Medicine, University of Oxford, John Radcliffe Hospital, Oxford

Professor Jonathan N Weber
Department of Genito-Urinary Medicine and Communicable Diseases, St Mary's Hospital Medical School, London

L Wolpert FRS
Professor of Biology as Applied to Medicine, Department of Anatomy, University College London, London

Preface

The purpose of *Basic Molecular and Cell Biology* was aptly described in a previous edition as allowing experts in the field to "explain the techniques of molecular and cell biology . . . and describe their implications for medicine". The response to the first two editions of this work indicates that they have amply fulfilled their aim and have allowed clinicians specialising in a range of different subjects to understand the manner in which the spectacular progress made in molecular and cellular biology in recent years is now contributing to the understanding, diagnosis, and ultimate therapy of human disease.

The continued rapid progress in this area now necessitates this new edition, in which the chapters in the preceding edition have either been extensively revised by their original authors or replaced by completely new chapters written by outstanding experts in the appropriate field. Most importantly, the opportunity has been taken to provide four additional chapters which deal with areas where rapid progress has been made in the past few years, such that a separate chapter on the topic is now required. These include, for example, programmed cell death (apoptosis) which, together with the process of cell division, regulates the number of cells in the body during development and adult life. Similarly, the inclusion of a chapter on gene therapy in the new edition serves to complement the pre-existing chapters on the therapeutic role of monoclonal antibodies and of other recombinant proteins.

In preparing this new edition the opportunity has also been taken of arranging the chapters in a more logical order, so as to allow the book to be used as a whole as well as being a collection of individual chapters. Thus after an introduction that provides an overview of the role of molecular and cell biology in clinical medicine, the next chapters present an overview of the methods used in molecular medicine, as well as a detailed account of the polymerase chain reaction which has had such a significant role in the progress made in recent years.

This methodological section is followed by a series of chapters dealing with the advances made in cell biology. These begin with an introductory chapter describing the nature of the cell which is followed by chapters dealing with the role of stem cells in growth and disease, as well as on the complementary topics of cell reproduction and apoptosis. The remaining chapters in this section deal with specific regions of the cell and are arranged in a manner allowing the reader to progress from the cell membrane to the nucleus. This section thus begins with a chapter on cellular adhesion and one on signalling by cell surface receptors. Subsequent chapters deal with the cell cytoplasm, transport across intracellular membranes, and the cytoskeleton. The remaining chapters in this section provide an overview of the cell nucleus and deal in detail with the process of gene regulation which occurs primarily within this organelle.

Having described in detail the insights obtained into the nature of the cell itself, the next section of this book deals with the insights that have been obtained by applying molecular biology to the study of particular diseases. Chapters in this section deal in detail with the role of specific genes in cancer and in the production of human congenital malformation. This is followed by a chapter dealing with the molecular genetics of common diseases which is obviously an important topic, not only in terms of our understanding of these diseases, but also in their diagnosis. The role of molecular biology in diagnosis is considered from a more clinical aspect in the next chapter which deals with the impact of molecular biology on clinical genetics and in particular on prenatal diagnosis.

Of course, ultimately insights obtained into the basis of particular diseases and their improved diagnosis must lead to a better therapy for patients who have such diseases. Thus the final section of this work discusses the prospects for improved therapy using monoclonal antibodies or recombinant proteins as well as the more recent progress made in the new field of gene therapy.

Overall, it is hoped therefore that this new edition will build on the success of its predecessors and, as was stated in the first edition, will continue "to give clinicians an insight into the way the medical sciences may be moving over the next few years and into the exciting possibilities opening up for the treatment of genetic disorders, cancer and the common illnesses of Western society".

David S Latchman

1: Molecular and cell biology in clinical medicine: introduction

David J Weatherall

In the nine years since this book was first published, there has been remarkable progress in the application of molecular and cell biology techniques to medical research. This new edition outlines some of these developments with particular emphasis on their application to the study of human disease. Although, as in previous editions, for convenience of presentation it falls into two parts, one covering molecular biology and the other cell biology, the two fields are inseparable; molecular biology describes the anatomy and organisation of the molecules of living organisms, and modern cell biology encompasses the way in which they work together as an orchestrated whole to mediate and regulate cellular function.

By way of an introduction to this important and timely collection, I shall summarise briefly the significance of some of the recent technical developments in this field, together with the clinical relevance of some of the ground that is covered.

Technical advances

Since the first edition of this book, there have been a number of seminal technical advances in molecular biology which have already made important contributions to human molecular genetics. Perhaps the most important, and one that merits a chapter of its own, is the polymerase chain reaction (PCR) (see chapter 3). This valuable technique makes it possible to amplify small amounts of DNA in a very short time and has formed the basis for a whole variety of new approaches which are particularly valuable for medical research. In particular it has been possible to use this method in the clinic for the rapid identification of genetic disease in fetal DNA (see chapter 17) and it has been modified to provide new and simpler approaches for DNA sequencing and identification of linkage markers.

Rapid progress has also been made in defining highly polymorphic regions of DNA including so called mini- and microsatellite DNA. This has been of particular value for identifying linkage markers, notably in mice, rats, and humans. Indeed, the generation of a linkage map of the whole of the human genome is well advanced and it is predicted, perhaps rather optimistically, that we may have a physical map of the entire human genome by the end of the second decade of the next century. With this in mind a great deal of effort is being put into developing automated methods for sequencing, including the use of robotics.

These new developments, which taken together constitute the Human Genome Project, should greatly facilitate the medical applications of recombinant DNA technology. Not only will it make it easier to find genes for monogenic disorders, it should greatly facilitate the identification of some of the important genes which are involved in the polygenic systems that underlie heart disease, hypertension, diabetes, and many more of our major killers.

Some applications of molecular biology in clinical medicine

The most immediate medical application of recombinant DNA technology has been in clinical genetics, in which it has been possible to work out the molecular pathology of many single gene diseases, and to institute new techniques for identification of carriers and for prenatal diagnosis. We now have a very good idea of the repertoire of mutations that underlie single gene disorders, although there are still some surprises. For example, the recent discovery of the cause of the fragile X syndrome and several other genetic disorders of the nervous system, which result from the amplification of short lengths of DNA either close to or within particular genes, underlines the extraordinary diversity of the molecular pathology of genetic disease.

In the first edition of this book, we predicted that the application of reverse genetics, that is, finding genes for disorders of unknown aetiology by genetic linkage, sequencing them, and then predicting the structure of the protein product of the mutant gene, would be an invaluable tool for human genetics. This has turned out to be the case. This approach, which has now changed its name to positional cloning, has been responsible for the discovery of many medically important genes, notably those that are involved in cystic fibrosis, muscular dystrophy, and polycystic kidney disease. The remarkable power of this field is evidenced by the extraordinarily rapid progress in understanding the pathology of cystic fibrosis which followed the discovery of the gene involved. Within two years over 100 different mutations had been discovered, the function of the gene product as a chloride channel had been defined, at least in outline, and thoughts are already turning to gene replacement therapy.

Although less spectacular, there has been steady progress towards an understanding of the molecular basis for the remarkable phenotypic diversity of single gene disorders. For example, different mutations at the β globin gene locus may give rise to a spectrum of disorders ranging from severe haemolytic anaemia, through congenital polycythaemia, to the diverse thalassaemia syndromes. Similarly, subtle differences in the site of mutations at the fibroblast growth factor receptor 2 locus may produce the widely differing phenotypes of Crouzon's and Apert's syndromes. As more is learned about gene regulation, it is becoming clear that many important single gene disorders associated with mental handicap and dysmorphology result from mutations at loci for a variety of *transacting* factors involved in the regulation of different families of genes.

Gene therapy, mentioned only briefly in earlier editions of this book, has made slow progress over the last nine years, and now merits a separate chapter (chapter 20). A number of retrovirus vectors have been constructed and methods have been worked out for ensuring the safety of this form of gene transfer. Other viral vectors have been developed and there has been some movement towards using nature's way of swapping genes, that is, by site directed recombination. There have, however, been few clinical successes with gene therapy and a recent report from the National Institutes of Health, Bethesda, USA underlines the importance of further basic research in this field, and the danger of premature experimentation with patients until more is known about the biology of gene transfer.

In introducing previous editions of this book, I predicted that it would be in the study of the common illnesses of Western society, such as degenerative vascular disease, diabetes, psychiatric disorders, cancer, and rheumatism, that molecular genetics might, in the long term, play its most important role. In particular, it is possible that, by defining some of the important genes in the polygenic systems that underlie our common diseases, we might learn more about their cause. There has been some genuine progress in research in this area although, not surprisingly perhaps, there have been false leads as well as successes. Perhaps the most spectacular advances have been made in understanding of the genetics of both type I (insulin dependent) and type II (non-insulin dependent) diabetes. Some of the major players in the genetics of vascular disease have started to emerge, but the very number of candidate genes for disorders of this type make it apparent that predictive genetics in this field is going to be extremely complex. Equally encouraging are the discoveries that common polymorphisms of several different genes may be involved in susceptibility to Alzheimer's disease, and that equally common polymorphisms of the vitamin D receptor gene may play an important role in determining the degree of bone density in postmenopausal women. Thus, there now seems little doubt that, as we come to understand the actions of these variant genes, we shall learn a great deal more about the pathogenesis of the common diseases of Western society (see chapter 16).

Perhaps the most spectacular progress towards an understanding of the pathogenesis of our common killers over the past few years has come from work in the cancer field. There have been major advances in understanding the role of mutations of oncogenes in the generation of cancer and the role of anti-oncogenes has been clearly defined (see chapter 14). Work on colon cancer has given some indication of the number of different mutations that may be required to generate a common malignancy, and recent studies on the genetics of breast cancer have defined at least two gene loci that are involved in familial forms of this common disease.

Cloned genes are already finding wide application in the diagnosis of bacterial, viral, and parasitic illnesses, and probes for oncogenes and immunoglobulin, and T cell receptor genes, are being used to type tumours. Several human genes have been persuaded to synthesise their products in bacterial systems. The correction of the anaemia of renal failure by genetically engineered erythropoietin is a good example of an early success in the rapidly expanding biotechnology industry. Numerous other therapeutic and diagnostic agents are under evaluation (see chapters 18 and 19). The problem of antibiotic resistance is being approached by producing hybrid antibiotics, constructed by altering the genes responsible for antibiotic production in fungi and bacteria. Recombinant DNA technology offers new approaches to vaccine production, for example, augmenting vaccinia virus with genes from other viruses such as rabies, and the production of large quantities of specific antigens from malarial parasites.

Amalgamation of molecular and cellular biology

Although less newsworthy, advances in our understanding of cell biology have been almost as rapid as those in molecular biology. This subject has equal potential for medical application, particularly for understanding cancer, autoimmune disease, human development, and the genesis of congenital malformation.

The application of recombinant DNA technology to the problems of cell biology is starting to unravel some of the mysteries of cell behaviour. For example, in many tissues there are pluripotential stem cells that go through cycles of multiplication before entering terminal differentiation towards their specialised functions (see chapter 5). It is becoming possible to define specific genes implicated in the regulation of different phases of the cell cycle, and in the critical steps of commitment and differentiation. The availability of gene probes and methods for studying the physical properties of inactive and active genes is enabling many of the processes of proliferation and differentiation to be described at the molecular level.

In understanding the orchestration of cellular proliferation and function, it is important to learn how cells respond to external regulatory signals such as hormones and growth factors. Considerable progress has been made in relating the action of external regulatory molecules to the protein kinase C messenger systems that are dependent on cyclic AMP, inositol triphosphate, and phospholipid. Equally important advances have been made in purifying some of the growth factors and other regulatory proteins that control cell proliferation. Once their structure is known their genes can be isolated and, by recombinant DNA technology, used to produce large quantities of their products for determining how they work and assessing their clinical potential. A whole family of haemopoietic growth factors has been obtained in this way, many of which are available for analysing how they regulate marrow progenitors and for use in clinical trials.

An understanding of what controls cell proliferation and differentiation has important applications for clinical practice. As mentioned earlier, the discovery of oncogenes, genes that are normally concerned with a variety of cellular housekeeping activities, and an understanding of their abnormal function in cancer have opened up a whole new area of investigation into the basic processes of malignant transformation. The finding that abnormal activation of these genes is often related to the chromosomal changes that are found in many common cancers has brought together cell biology and cytogenetics in an exciting new area of cancer research. The message from these studies is that cancer is a diverse disorder resulting from genetic damage of many kinds, including recessive or dominant mutations, major DNA rearrangements, and point mutations, all of which may disturb the expression of biochemical function of the affected genes. We now have the tools with which to unravel these processes.

Studies of cell membrane and receptor function promise to have equally important clinical applications. For example, work on the lifestyle of the low density lipoprotein receptor, including its internalisation and recycling and how these are regulated, has helped to define the many stages at which cholesterol metabolism might be altered and has offered us new possibilities for the pharmacological control of cholesterol. By means of monoclonal antibody and protein engineering technology, it will probably be possible to modify receptor function. The isolation of the T cell receptor and studies of the cell biology of antigen recognition have opened up a completely new approach to sorting out the mechanisms of the autoimmune processes that may underlie many common diseases. The discovery and characterisation of the "supergene" family that regulates the immune system and the lymphokines that constitute its cellular regulatory molecules are another remarkable example of the combined strengths of cell and molecular biology and their potential for medical practice.

Another rapidly growing area of the amalgamation of molecular and cell biology is in the study of the mechanisms of ageing. Work in yeast and other model organisms has provided data which suggest that there are genes that are involved in altering the lifespan of particular organisms. There is increasing evidence that the products of at least some of these genes are involved in protection of DNA against endogenous oxidant damage. There has also been considerable progress towards an understanding of the complex changes that occur in the structure of proteins during the ageing process. These observations promise to be of considerable importance, not just in the field of ageing but also in helping us to understand some of the pathological processes that are associated with old age, notably vascular disease, cataract, dementia, and late onset diabetes.

The future

It is hoped that this collection of short articles will give clinicians some insight into the way in which the medical sciences may be moving over the next few years. In effect, we appear to be changing from the era of whole patient physiology and pathology to the study of disease at the molecular and cellular level. This will undoubtedly alter the pattern of medical research. One effect will be to cloud the current compartmentalisation of medicine into watertight specialties; cardiologists, nephrologists, and so on are now using the same techniques of cell and molecular biology to study diseases particular to their subject. It therefore follows that we may have to revise the way that research departments and resources are organised if we are to take full advantage of what recombinant DNA technology has to offer.

Clinicians may be worried about this reductionist approach to the study of disease, particularly at a time when they are being accused of ignoring the broader pastoral aspects of patient care and when there is a cry for a return to holistic medicine—that is, to rediscover how to be a good doctor. But there is no reason why a proper understanding of disease processes at the molecular and cellular level should have a deleterious effect on clinical care or on doctors' attitudes to their patients. On the contrary, in the long term it should enable us to treat many diseases more rationally and hence to reduce the proliferation of high technology medicine that characterises much of the current medical scene. This will not, however, happen overnight; there is likely to be a long period of development and evaluation of our new technology before its clinical application can be fully assessed. Indeed, once we have dissected the human genome and know more about the products of the hundred thousand or so genes that constitute it, it may be necessary to develop more sophisticated techniques of biomathematics to try to develop an integrative physiology, that is, to find how the whole thing works as an orchestrated whole. It may well require this type of approach before the full potential of the molecular era of clinical research is fulfilled.

All this will require some rethinking about medical education and the way in which our clinical schools and postgraduate programmes are organised in the future; the profession will have to take on board and accommodate individuals with a much broader range of scientific skills than has hitherto been necessary. This will require some urgent thought by those who are concerned with training our doctors in the future. I hope that this book will help to provide some background information on which we can develop an informed debate on how these changes might be made while, at the same time, maintaining high standards of patient care.

2: Methods in molecular medicine

Paul M Brickell

The chapters in this book illustrate the many ways in which molecular biological techniques have revolutionised our understanding of normal and abnormal cellular processes, and have allowed the introduction of powerful new diagnostic and therapeutic approaches for a range of human diseases. It will become clear in the pages of the book that this success has involved the collaboration of molecular biologists, biological scientists, clinicians, primary health care specialists, and many other groups of workers. The purpose of this chapter is to provide a general background for what follows, by outlining the steps involved in gene expression and by describing some of the techniques commonly used to investigate gene structure and expression in both health and disease.

Gene expression

A DNA molecule consists of two strands wound around each other in a double helix. Each strand has a backbone of repeating sugar–phosphate units, and attached to each sugar residue is one of four bases, named adenine (A), cytosine (C), guanine (G), and thymine (T). The two strands are held together by hydrogen bonding between bases, so that A binds only to T, whereas C binds only to G. The sequences of the bases on the two strands are said to be complementary.[1]

One complete set of human genetic information comprises about 3×10^9 base pairs (bp) of double stranded DNA. A diploid cell contains two copies of this genome, distributed between 23 pairs of chromosomes. Some regions of the genome encode proteins and others contain genetic information for the synthesis of ribosomal RNA and transfer RNA molecules, which form part of the cell's machinery for protein synthesis. These regions are termed "genes". The great majority of the genome does not contain genes, however, and although some of this non-coding DNA consists of unique DNA sequences, much of it comprises DNA sequences that are repeated thousands, and in some cases millions, of times.[1]

The steps involved in the expression of eukaryotic protein coding genes are outlined in figure 2.1. One strand of the DNA is first

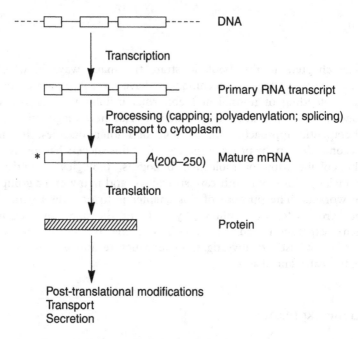

FIG 2.1—*Steps in the expression of a typical eukaryotic protein coding gene. Exons are shown as white boxes. *5′ cap.*

copied into single stranded RNA by the enzyme RNA polymerase II. This process is termed "transcription". Many genes encode proteins that are required in all cell types, and these genes are transcribed in all cells. Other genes encode proteins that are required only in certain cell types, only at certain times during the organism's development, or only in response to certain stimuli. Most genes of this type are transcribed only under certain circumstances. The regulation of transcription is highly complex and involves the interaction of many different cellular proteins with RNA polymerase II and with DNA sequences adjacent to transcribed genes.[12]

The primary RNA transcript of a gene undergoes three types of modification, which convert it into mature messenger RNA (mRNA). First, the start of the RNA molecule (the 5' end) is chemically modified, or capped, to enable it to interact with the cellular machinery for protein synthesis. Second, the other end of the molecule (the 3' end) is trimmed back and about 200–250 adenine residues are added. This is termed "polyadenylation", and the resulting poly(A) tail promotes RNA stability. One the most startling of the discoveries to follow the cloning of eukaryotic genes in the 1970s was the finding that eukaryotic primary transcripts, unlike those of prokaryotes, do not code directly for protein. Rather, the protein coding regions, or exons, of a eukaryotic gene and its primary transcript are separated by intervening sequences, or introns, which do not code for amino acid sequences found in the translated protein. The third modification to the primary transcript is therefore to remove the intervening sequences by a process called splicing, which joins the exon sequences together.[12] In some cases the primary transcripts of a gene are spliced differently in different cell types to yield mRNA molecules encoding different proteins.[12]

The mature mRNA is transported from the cell nucleus to the cytoplasm where it is translated into protein. During translation, the sequence of bases in the mRNA is decoded in groups of three, each triplet codon specifying the addition of a particular amino acid to the growing polypeptide chain.[1] The newly formed polypeptide chain folds and may undergo post-translational modifications such as proteolytic cleavage, glycosylation, or phosphorylation. The protein may also be transported to a particular subcellular compartment, such as the nucleus or the surface membrane, or may be secreted from the cell.[3]

Cloning gene sequences

The steps involved in cloning fragments of DNA are illustrated in figure 2.2, and further details may be found elsewhere.[4] To clone single gene sequences from the enormous amount of genomic DNA in each human cell, the DNA is first cut into smaller fragments using restriction endonucleases.[4] These are enzymes that bind to specific short sequences of base pairs and cut the backbone

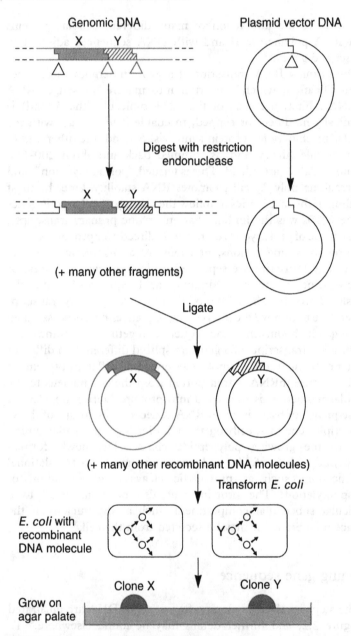

FIG 2.2—*Steps involved in cloning DNA fragments into a plasmid vector. White arrow heads indicate restriction endonuclease sites.*

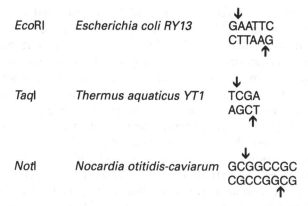

FIG 2.3—*Some examples of restriction endonucleases. The organisms from which the enzymes are isolated and the recognition sites are shown. Arrows indicate the positions at which the backbones of the DNA strands are cleaved.*

of the two DNA strands (fig 2.3). The fragments can then be isolated from each other and produced in large quantities by means of DNA cloning. This is achieved by ligating, or joining, each of the genomic DNA fragments to a vector DNA molecule, which has the ability to replicate in the bacterium *Escherichia coli*. The simplest vectors are those based on bacterial plasmids, which are naturally occurring circular DNA molecules that replicate in *E. coli*, and figure 2.2 illustrates the use of a plasmid vector. The recombinant DNA molecules produced by ligation are put into *E. coli* cells by a process termed "transformation", so that transformed cells each receive only one recombinant DNA molecule. This will be replicated to a high copy number so that the progeny of a single transformed cell will each contain many copies of the same recombinant DNA molecule. When grown on an agar plate, each transformed cell gives rise to a bacterial colony containing many copies of a single cloned human genomic DNA fragment. This is a genomic clone. If this collection of cloned fragments is sufficiently large, the bacterial colonies will contain, between them, every fragment in the genome. Such a collection of cloned fragments is called a genomic library.

In practice, plasmid vectors are not particularly suited to building libraries of large genomes such as the human genome, not least because they will only comfortably accept inserted fragments of up to about 10 000 bp, which is smaller than many human genes.

Vectors based on bacteriophage λ, which is a bacterial virus, will accept fragments of about 20 000 bp, and are also more efficient tools for constructing libraries of large genomes. The current vectors of choice are, however, cosmids, which combine features of plasmids and bacteriophage λ and can accept inserts of 30 000–40 000 bp, or yeast artificial chromosome (YAC) vectors, which can accept inserts of up to several million base pairs.[4] A typical human genomic library constructed in a cosmid vector would comprise about 2000–500 000 clones.

It is not possible to clone single stranded mRNA molecules. An enzyme called reverse transcriptase can, however, be used to convert mRNA into double stranded complementary DNA (cDNA) in vitro,[4] and the double stranded cDNA molecules can then be cloned, essentially as described above. A collection of clones, which between them contain cDNA copies of every mRNA molecule in a cell, is called a cDNA library and typically contains at least 200 000 clones.

Having generated a genomic or cDNA library, the next step is to search through the library for the clone containing the particular gene or mRNA sequence of interest. With so many colonies to search through, this process of library screening can be a considerable undertaking. There are, however, a number of well established library screening procedures, descriptions of which may be found elsewhere.[4] Having isolated the desired genomic or cDNA clone, it can be used to analyse the structure and expression of the corresponding human gene.

Nucleic acid hybridisation

Many analytical techniques depend on nucleic acid hybridisation.[5] When double stranded DNA molecules in solution are heated or treated with alkali, the hydrogen bonds between the bases break and the two strands separate. This process is termed "denaturation". If the solution is then allowed to cool slowly, or if its pH is lowered, complementary bases pair with each other and double stranded DNA molecules, indistinguishable from the starting material, re-form. This is termed "renaturation" or "annealing". As long as they have complementary sequences of bases, DNA strands from completely different sources will also renature in this way, forming DNA–DNA hybrids. DNA–RNA and RNA–RNA hybrids can also be formed between complementary RNA and DNA strands. Thus, if a single stranded nucleic acid molecule is labelled with a radioactive or non-radioactive tracer, it can be used as a probe to search for complementary nucleic acid strands, to which it will hybridise. If the probe corresponds to the sequence of a gene, then it can be used to identify that gene, or the mRNA transcribed from it, in whole preparations of DNA or RNA isolated from cells and tissues.

Analysing gene structure

Southern blotting

The structure of a gene can be analysed using the technique of Southern blotting, which was discovered by Professor EM Southern. The steps involved in this technique are shown in figure 2.4, along with a typical result. Genomic DNA can be isolated from any tissue, but peripheral blood leucocytes are a convenient source in humans. The DNA is digested with a restriction endonuclease, generating a large number of DNA fragments of differing sizes. These can be separated according to their size by agarose gel electrophoresis: the smaller the fragment, the further it will migrate in an agarose gel under the influence of the electrical field.[5] To permit subsequent hybridisation to a gene probe, the double stranded genomic DNA fragments in the gel are denatured by soaking the gel in alkali. Attempts to hybridise a labelled gene probe directly to the fragile gel would cause it to disintegrate, and so the fragments are transferred from the gel to a nylon membrane by a capillary transfer procedure, which provides a "hard copy" of the denatured genomic fragments.[5] The membrane is then incubated with a single stranded labelled probe for the gene of interest, under conditions that allow the probe to hybridise with complementary gene sequences on the membrane. Excess probe is washed off and the position of bound probe is detected by autoradiography with x ray film, in the case of radiolabelled probes, or by an appropriate colorimetric detection method for non-radioactive probes. The genomic fragment containing the gene sequence appears as a black band on the film (fig 2.4) or a coloured band on the membrane, respectively. The sizes of fragments, in base pairs, can be determined by reference to the migration of DNA fragments of known size.

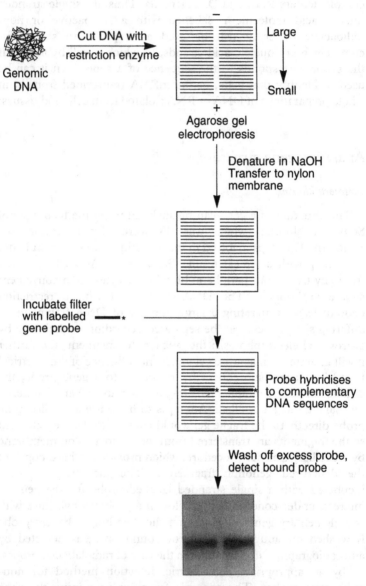

FIG 2.4—*Southern blotting and hybridisation: genomic DNA from two people was digested with the same restriction endonuclease and hybridised with the same gene probe.*

Restriction fragment length polymorphisms

In figure 2.4, the probe hybridised to a DNA fragment of equal size in two people. This is the most common result of such an experiment. Occasionally, however, a probe hybridises to fragments of different sizes in different individuals, for example, figure 2.5

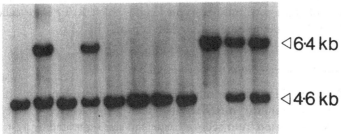

FIG 2.5—*Detection of an RFLP by Southern blotting and hybridisation. Genomic DNA from 11 unrelated healthy people was cut with TaqI and hybridised with a radiolabelled probe for human 5-HT$_{1a}$ (serotonin) receptor gene sequences. Sizes of fragments are given in kilobase pairs.*

shows the result of hybridising a Southern blot of *Taq*I digested genomic DNA from 11 unrelated, healthy people with a probe for the human serotonin receptor gene sequence. In the human population, this sequence may lie on either a 4600 bp (4·6 kilobase pair or 4·6 kb) *Taq*I fragment or a 6·4 kb *Taq*I fragment. Individuals 1, 3, 5, 6, 7, and 8 are homozygous for the smaller fragment, individual 9 is homozygous for the larger fragment, and individuals 2, 4, 10, and 11 are heterozygotes. This probe therefore detects a genetic difference between healthy individuals. Such a difference is termed a "polymorphism" and as the polymorphism is detected as a difference in restriction fragment lengths, it is termed a "restriction fragment length polymorphism" or RFLP. There is nothing mysterious about an RFLP; it is simply a marker of a genetic difference between healthy individuals, such as eye colour, but detected by a slightly more sophisticated technique. As will be seen elsewhere in this volume, the existence of RFLPs allows mapping of DNA sequences, tracking of mutant genes through families, and prenatal detection of inherited diseases.

Restriction fragment length polymorphisms commonly arise in two ways.[6] First, a change in a single base pair may create or destroy a restriction enzyme site, which is consequently present in some individuals but not in others. This will lead to the generation of restriction fragments of unequal length. Second, an extra piece of DNA may be inserted into, or deleted from, the DNA of some individuals. This commonly arises from the presence of repeated DNA sequences which are scattered throughout the human genome in blocks containing multiple copies of the same sequence. The number of copies of the repeat in a given block can vary from one individual to another, so that a restriction fragment containing the block of repeats will also vary in length. Such RFLPs are called VNTR (variable number of tandem repeats) RFLPs.

Northern blotting and in situ hybridisation

Northern blotting is a modification of Southern blotting which enables detection of mRNA rather than genomic DNA.[5] Briefly, mRNA is isolated from cells or tissue and size separated by agarose gel electrophoresis. As mRNA molecules are single stranded, they form weak intra- and intermolecular hydrogen bonds. The resultant folding or aggregation of mRNA molecules means that they do not migrate true to size. Agarose gels for electrophoresis of mRNA therefore contain formaldehyde as a mild denaturing agent. The electrophoresed mRNA is transferred to a nylon membrane as in Southern blotting, and the membrane is then incubated with probe, washed, and the hybridised probe detected as before. Figure 2.6A shows a northern blot of mRNA isolated from chick embryo liver and trunk (excluding liver) which has been hybridised with a probe for mRNA encoding the chicken RXR-γ (retinoid-X-receptor-γ) nuclear retinoid receptor.[7] Such an experiment shows the size of the mRNA and the quantities present in tissues, but does not show which cells within a given tissue contain it. Thus, RXR-γ mRNA could be present at low levels in all cells within the trunk, or could be expressed at high levels in some cells while being absent from others. To address such problems, the in situ hybridisation technique was developed,[8] in which a radioactive or non-radioactive probe is hybridised to RNA within cells in tissue sections. As shown in figure 2.6B, in situ hybridisation of the chicken RXR-γ probe to a section of chick embryo trunk reveals that RXR-γ RNA is restricted to the peripheral nervous system. This information is clearly of value in determining the function of the RXR-γ protein. More recently, this technique has been adapted to allow hybridisation of non-radioactive probes to RNA in whole pieces of tissue or whole embryos (fig 2.7). This technique, called whole

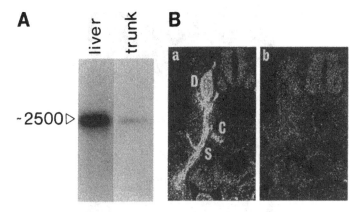

FIG 2.6—*Detection of mRNA: (A) northern blot of mRNA isolated from embryonic chick liver or trunk (excluding liver) was hybridised with a radiolabelled probe for chicken RXR-γ mRNA.*[8] *The size of the mRNA is shown in bases. (B) In situ hybridisation of adjacent transverse sections through a chick embryo using a radiolabelled probe for chicken RXR-γ RNA (a) or a radiolabelled negative control probe (b). These are dark field photographs in which hybridisation signal appears as white grains. D, dorsal root ganglion; S, spinal nerve; C, sympathetic chain.*

mount in situ hybridisation,[7] shows patterns of gene expression in three dimensions, which can be difficult to reconstruct from in situ hybridisation to serial tissue sections.

Each of these techniques can give us some measure of the steady state level of an mRNA species in a particular cell type. A common error is to equate this with transcription rate, but in fact the steady state level of an mRNA species represents a balance between its synthesis (transcription) and its degradation. To measure the transcription rate of a gene directly, a nuclear run on assay[2] can be used. The stability of a particular mRNA can be measured by an adaptation of the northern blotting technique.[5]

Western blotting

As described above, Southern blotting is used to detect specific DNA sequences, whereas northern blotting is used to detect specific mRNAs. Western blotting allows detection of specific proteins. Proteins in cell extracts are size separated by sodium dodecyl-sulphate–polycrylamide gel electrophoresis and transferred electrophoretically to a nitrocellulose membrane. The membrane is then incubated with antibodies against the protein of interest

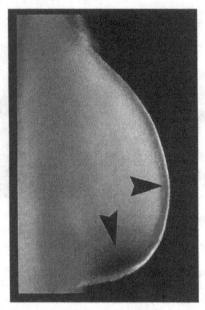

FIG 2.7—*In situ hybridisation to a whole mount preparation of a chick embryo wing bud using a non-radioactive probe for chicken bone morphogenetic protein-2 (BMP-2) RNA.*[9] *Hybridising cells stain purple. In this photograph, hybridising regions appear dark and are indicated by arrows.*

and bound antibodies are detected by an enzyme linked assay.[10] As with northern blotting, this technique gives information about the size and quantity of the molecule detected. The spatial distribution of protein in tissues can be determined by immunohistochemistry, in which antibodies are bound to proteins within tissue sections.[11]

The polymerase chain reaction

The polymerase chain reaction (PCR) was developed much later than most of the techniques described above, and has rapidly found a wide range of applications.[4 12] It allows the production of large quantities of a specific region of DNA without recourse to cloning. All that is required is a knowledge of short lengths of the base pair sequences on either side of the region to be amplified. The polymerase chain reaction is fast, relatively simple to perform, and requires only very small quantities of genomic DNA. At the limit, specific DNA sequences can be amplified by PCR from single cells.[13] The steps in a typical PCR are shown in figure 2.8. Short

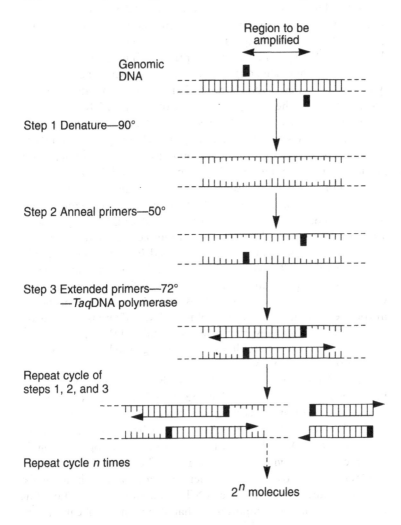

FIG 2.8—*The polymerase chain reaction: the temperature of the annealing step (step 2) will vary depending on the sequence of the oligonucleotide primers.*

single stranded DNA molecules, which are complementary to sequences flanking the region of DNA to be amplified, are chemically synthesised. These oligonucleotides, which are typically

20–25 bases long, are mixed in solution with genomic DNA, *Taq* DNA polymerase, and the subunits for DNA synthesis. The temperature of the mixture is raised to 90°C to denature the genomic DNA and then lowered to 50°C so that the oligonucleotides anneal to the complementary sequences in the genomic DNA. The temperature is then raised to 70°C to allow the *Taq* DNA polymerase to work optimally. The enzyme binds to the ends of the annealed oligonucleotides as shown and these prime synthesis of a new DNA strand that is complementary to the existing strand. These three steps constitute one PCR cycle. The same cycle is repeated 30–40 times and, as the number of amplified molecules doubles at each cycle, this yields a large quantity of DNA, all of which corresponds to the region between the oligonucleotide primers. The amplified DNA can be visualised by agarose gel electrophoresis. The repetitive cycles are usually performed by placing the reaction tube in a metal heating block, the temperature of which is controlled by a microprocessor. Most DNA polymerases are unstable at the high temperature required for the denaturation step, but *Taq* DNA polymerase is isolated from the thermophilic micro-organism *Thermus aquaticus* and so remains active throughout the procedure. Early versions of *Taq* DNA polymerase were rather slow at copying DNA strands and were prone to introduce mutations into the amplified DNA. There is now a wide variety of commercially available, heat stable DNA polymerases which overcome these problems, allowing high fidelity amplification of long regions of DNA.[12]

The polymerase chain reaction can be used to identify DNA polymorphisms, as illustrated in figure 2.9. In this example, the oligonucleotide primers correspond to sequences that flank the site of the CA repeat block. Every individual has multiple copies of the sequence CA at this site, but the number of copies is highly variable from individual to individual. The length of the DNA fragment amplified during PCR therefore varies among individuals, and is another example of a VNTR polymorphism. The PCR technique is of such importance that it and its applications are described in more detail in chapter 3.

Conclusion

This chapter has outlined a few of the most common techniques in the molecular biologist's armoury. Powerful though they are,

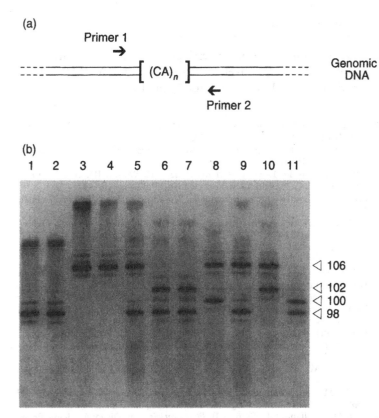

FIG 2.9—*Detection of a polymorphism by PCR: (a) the oligonucleotide primers flank a block of CA repeats. (b) Genomic DNA from 11 unrelated healthy people was amplified by PCR using the primers shown in (a). The sizes of the major amplified products are indicated in base pairs. Individuals 1, 2, 3, and 4 are homozygotes; the rest are heterozygotes. The minor bands in each track represent background and should be ignored.*

they are continually being updated and refined, and many of their refinements will be discussed in other chapters in this book. It is now some 20 years since the first genes were cloned and blotting techniques were developed, whilst PCR is almost 10 years old. In this time, techniques that were once seen as the province of a few specialised laboratories now have a routine place in research and hospital service laboratories throughout the world. As will be seen throughout this book, their application continues to allow remarkable progress to be made in our understanding of gene structure, expression, and function in health and disease.

Acknowledgments

I am grateful to Robin Sherrington of the Department of Academic Psychiatry, University College London Medical School for providing data shown in figures 2.5 and 2.9.

1 Darnell J, Lodish H, Baltimore D, eds. *Molecular cell biology*, 2nd edn. New York: Scientific American Books, 1990.
2 Latchman DS. *Gene regulation: a eukaryotic perspective*, 2nd edn. London: Chapman & Hall, 1995.
3 Alberts B, Bray D, Lewis J, Raff M, Roberts K, Watson JD, eds. *Molecular biology of the cell*, 3rd edn. New York: Garland Publishing, 1994.
4 Brown TA. *Gene cloning: an introduction*, 3rd edn. London: Chapman & Hall, 1995.
5 Darling DC, Brickell PM. *Nucleic acid blotting: the basics*. Oxford: Oxford University Press, 1994.
6 Davies KE, Read AP. *Molecular basis of inherited disease*. Oxford: IRL Press, 1988.
7 Wilkinson DG, ed. *In situ hybridization: a practical approach*. Oxford: Oxford University Press, 1993.
8 Rowe A, Eager NSC, Brickell PM. A member of the RXR nuclear receptor family is expressed in neural-crest-derived cells of the developing chick peripheral nervous system. *Development* 1991;**111**:771–8.
9 Francis PH, Richardson MK, Brickell PM, Tickle C. Bone morphogenetic proteins and a signalling pathway that controls patterning in the developing chick limb. *Development* 1994;**120**:209–18.
10 Gershoni JM, Palade GE. Protein blotting: principles and applications. *Analyt Biochem* 1983;**131**:1–15.
11 Polak JM, van Noorden S, eds. *Immunocytochemistry: modern methods and applications*. Bristol: John Wright, 1986.
12 Innis MA, Gelfand, DH, Sninsky JJ. *PCR strategies*. San Diego: Academic Press, 1995.
13 Handyside AH, Kontogianni FH, Hardy K, Winston RMI. Pregnancies from biopsied human preimplantation embryos sexed by Y-specific DNA amplification. *Nature* 1990;**344**:768–70.

3: The polymerase chain reaction: a tool for molecular medicine

A F Markham

The polymerase chain reaction has been unquestionably unique, as new techniques go, in the speed with which it has been embraced by non-experts in most specialties of the biological sciences, including medicine. The reason for this is the unusual simplicity of the procedure. In terms of its power to drive biological research, the advent of the polymerase chain reaction can certainly be compared with the discovery of the techniques of molecular cloning some 25 years ago. However, whereas years of training and practice were usually needed to master the many and complex skills of recombinant DNA technology, the complete beginner can start to perform polymerase chain reaction (PCR) experiments and generate meaningful results within a few days at most—hence the explosion of activity.[1-4]

The technique was first described in its initial format in 1985,[5] and over the next three years appreciation of its potential gradually became widespread. This potential was fully realised in about 1988. It coincided with the commercial development of two key components for the polymerase chain reaction: a DNA polymerase that could be heated at quite high temperatures (boiling water) without losing its activity, and robust machines that would quickly heat and cool samples repeatedly in a cyclic fashion.[6] Synthesis of the oligonucleotides required as primers in the reaction had already become a commonplace procedure. The reaction has become probably the most widely used single technique in all branches of the biological sciences.

What is the polymerase chain reaction?

The polymerase chain reaction is a delightfully simple concept, first alluded to 30 years ago,[7] and very reminiscent of the way that cells duplicate their DNA to expand their numbers in vivo. The other simple analogy is to the chain reaction of nuclear physics. The technique permits the analysis of nucleic acids (DNA or RNA) from any source.

Procedure

A small sample of DNA in solution is placed in a single tube. (As will become clear below, any RNA samples for analysis are simply converted to DNA in a single preliminary step.) Two oligonucleotides (which are easy to make artificially on automatic machines or can be purchased from several suppliers) are added to the tube. Their sequences are chosen so that they match two short DNA sequences that flank the region of interest. The exponential increase in the amount of DNA sequence that is now known, and the universal accessibility of these data through computer databases, mean that any scientist can quickly start to study genes of interest in his or her own laboratory. There is no need to obtain DNA clones from other workers, so all the problems previously associated with that are avoided.

Considerably more oligonucleotides than the DNA to be analysed are provided deliberately. A thermostable DNA polymerase is added to the same tube.[8] Deoxynucleoside triphosphates (dNTPs), and salts and buffer to allow the enzyme to work properly, are also included. All these reagents are available commercially. Taq DNA polymerase was isolated from the thermophilic organism, Thermus aquaticus. Not only can it tolerate heating at 100°C, but it makes DNA at high temperatures compared with the 37°C physiological temperature optimum of most enzymes. A cloned version of this enzyme without any nuclease activity is also available as are enzymes such as the PFu polymerase which has a proofreading function and can be used where the sequence of the PCR product may be determined accurately.

This simple mixture is placed in a heater and the temperature quickly raised to just below the boiling point of water. This causes the double stranded DNA in the sample to dissociate into two single strands as the hydrogen bonds, which hold the two strands together under physiological conditions, break down reversibly on heating. After a minute or so, the solution is allowed to cool towards physiological temperature. This allows hydrogen bonds to re-form. The two DNA strands in the sample would, of course, usually relocate their partners and re-form the paired double helix. However, in the PCR tube the oligonucleotides, which are present in great excess, quickly and highly specifically bind to their complementary single strands from the denatured sample. As soon as this happens the oligonucleotide can act as a primer for DNA polymerase and is extended to form a new double stranded molecule. Thus each double stranded DNA molecule in the original sample has been melted to form two single stranded molecules, which have then been turned into two double stranded molecules (see the box).

Outline of the procedure

- A small sample of DNA is placed in a tube
- Two oligonucleotides are added. These have sequences matching two sequences of the DNA that flank the region of interest
- A thermostable DNA polymerase is added
- The mixture is heated to just below 100°C and the DNA dissociates into two single strands
- The solution is allowed to cool and the single strands bind to the oligonucleotides, which are in excess
- The oligonucleotide now acts as a primer for DNA polymerase and is extended to form a new double stranded molecule
- The cycle is repeated, with the amount of DNA doubling each time

---

There is now twice as much double stranded sample DNA present in the tube as there was to start with. The cycle is then repeated and the amount of sample DNA doubled further with every cycle. This geometrical amplification is perfectly analogous to a nuclear chain reaction with twofold, fourfold, eightfold, sixteenfold, thirty twofold, etc. amplification at subsequent steps. After n cycles the degree of amplification is of course 2^n. Thus after 10 cycles 1024-fold amplification is achieved and after 20 cycles 10^6-fold amplification results.

No great manual dexterity is required to perform the technique, in that all these reactions go on by simply automatically heating and cooling without ever opening the reaction tube. The whole process is outlined in figure 3.1.

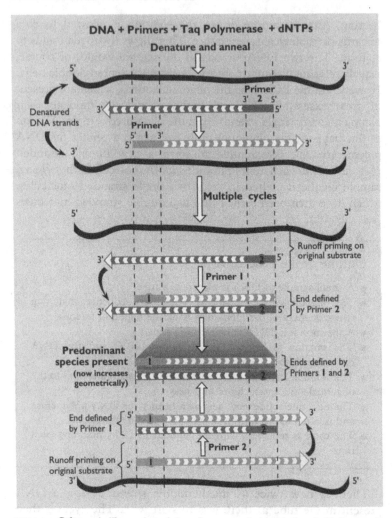

FIG 3.1—*Schematic illustration of the polymerase chain reaction. Initially runoff priming occurs, with the primers being extended to the end of the template. In subsequent cycles the product with ends defined by both primers 1 and 2 becomes predominant. It increases geometrically in amount by doubling with each cycle. After 30 cycles there are a billion copies of the reaction product but only 30 copies of the runoff products.*

The orientation of the two strands in double stranded DNA is important to note. They are described as being antiparallel, which means that when bound together in a helix one reads 5′ to 3′, the other 3′ to 5′. This simple consideration dictates the design of the synthetic oligonucleotides. When making double stranded DNA, *Taq* polymerase attaches new residues to the 3′ end of primers. Thus in the polymerase chain reaction DNA is essentially copied only between the two primers. After multiple cycles of amplification, the predominant double stranded DNA species in the sample is a fragment whose two ends are defined by the two primer oligonucleotides—that is, not all the DNA in the original sample is amplified. For example, if the sample DNA were total human DNA from a blood sample, and therefore contained 3×10^9 base pairs, amplification of any 300 base pair segment could be specifically achieved. It will be appreciated that the signal to noise problems inherent in detecting and analysing 300 base pairs out of 3 billion (1 in 10^7) are very much alleviated after 30 cycles of polymerase chain reaction, when 10^9 copies of the 300 base pair fragment are present for every single copy of the total genomic DNA background.

Problems

The polymerase chain reaction is not quite infinitely powerful or infallible. Eventually, after many cycles, the concentration of oligonucleotide primers falls because they have all been incorporated into products of the reaction. Similarly, the concentration of deoxynucleoside triphosphate substrates, and indeed the activity of the DNA polymerase itself, decline. Products of pyrophosphate breakdown may inhibit further reaction. Amplification will usually be sufficient for most analytical purposes by this stage, but should even greater sensitivity be necessary—for example, in the analysis of DNA from a single cell—then a tiny aliquot from the first PCR tube (as a source of sample DNA) is simply transferred to a second identical tube, and amplification is continued. Should specificity be a problem (for instance, if the target DNA comes in a sample that also contains many other closely related DNA sequences) then this can be overcome using "nested" primers in the second tube. These match sequences just inside the two original primer sites. Thus nested amplification of a product of the polymerase chain reaction depends overall on accurate recognition by four independent oligonucleotides.

The other aspect that may cause problems is the specificity of the priming reaction itself. This depends on several considerations. The size of the oligonucleotide will determine whether it occurs more than once in a sample DNA and therefore might prime

DNA polymerase activity at multiple sites. This will generate only spurious products of the polymerase chain reaction when unwanted priming also occurs close by on the other DNA strand. However, the reaction may be inefficient if one of the primers is depleted because of excessive spurious priming. Several simple measures can be taken to eliminate this type of problem should it arise. These include using raised temperatures during cycling, adjusting the magnesium concentration (which partly determines the ease of hybridisation), and various other easy tricks to destabilise partially mismatched primers. On balance, these considerations do not usually constitute a serious problem, and any teething troubles can usually be overcome quickly by minor and obvious adjustment to reaction conditions.

Why has the reaction had such a major impact?

Southern blotting

The impact of the polymerase chain reaction can be attributed to its practical simplicity and its speed, sensitivity, and specificity. Consideration of some of the routine day to day techniques of molecular biology allows this to be illustrated. In a genomic Southern blot it is impractical to load more than 10 μg (10^{-6} g) of restriction enzyme digested DNA per lane of an agarose gel. Higher loading causes overloading and streaking. If one is interested in a 3 kilobase fragment then this constitutes about 10 pg (10^{-12} g) of the sample. Detecting 10 pg of anything is taxing, and doing so when it is effectively "contaminated" with a millionfold excess of somewhat similar DNA fragments is even harder. The practical consequence has previously been that prolonged autoradiographic techniques were required. Days or weeks would be needed for each experiment. By simply amplifying the fragment of interest it can be characterised in agarose gels, usually by visual inspection, in a few hours.

Messenger RNA

Another example of the reaction's impact would be the use of the polymerase chain reaction to overcome some of the many problems inherent in analysing messenger RNA (mRNA) by classic techniques. These molecules, the ultimate source of information about what is going on at a specific time in a particular cell, are extremely labile chemically (for example, to traces of alkaline detergent in less than scrupulously clean glassware) and enzymatically (to the ubiquitous ribonuclease enzymes). A typical RNA preparation from a tissue sample or cell culture will contain around 2% of mRNA or less, with 98% ribosomal RNA (rRNA) and transfer RNA (tRNA). Though selecting for the mRNA by means of its tail of multiple A residues on oligo(dT) affinity columns is possible, it is notoriously difficult even for the nimble fingered. Handling the minute amounts of material required remains extraordinarily difficult. Synthesis and subsequent cloning of copy DNA (cDNA) often yielded unsatisfactory clone banks with short inserts and heavy ribosomal DNA (rDNA) contamination for these purely technical reasons.

The situation is transformed by the use of the RNA polymerase chain reaction (fig 3.2). Oligo(dT) itself can be used to prime first strand synthesis of cDNA. Commercial kits are available and even thermostable "reverse transcriptases" have been introduced. Double stranded cDNA is obtained by a number of methods including using a second oligonucleotide[9] or degenerate mixture specific for a target mRNA, or else total double stranded cDNA is produced by "tailing" the first strand by adding multiple copies of the same nucleotide, for example, homo(G), and using the complementary homo(oligonucleotide) as the second primer.[10] This has become known as an "anchor polymerase chain reaction". After multiple cycles of amplification, contaminating RNA species are no longer a problem and there is plenty of material to clone, sequence, or otherwise analyse, all within a few hours and with minimal manual intervention.

Incorporating the reaction into routine molecular genetics

Analysis of restriction fragment length polymorphisms

Analysis of restriction fragment length polymorphisms (RFLPs), previously a stock-in-trade of molecular genetics, has been revolutionised by the polymerase chain reaction. We have discussed the inherent difficulty of Southern blotting above. For RFLP analysis it is essential, in particular, to generate DNA of sufficient quality to undergo digestion to completion with restriction enzymes.

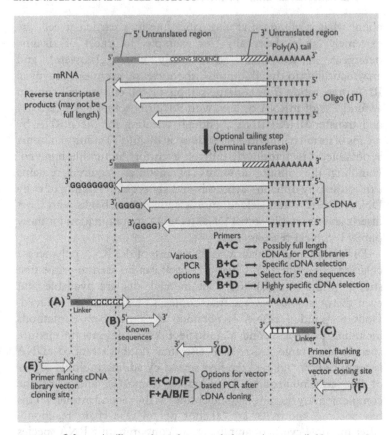

FIG 3.2—*Schematic illustration of some of the options available to analyse mRNA by the reverse transcriptase polymerase chain reaction. When single stranded cDNA has been synthesised, a variety of products can be generated by selecting pairs of the primer types A–F. As illustrated for primers A and C, linker sequences may be included at the 5' ends of any of these primers to further facilitate cloning of the resulting reaction products.*

This is not always easy. Failure to cut sample DNA may lead to diagnostic errors with clinical material. Amplifying the target DNA means that obtaining high quality DNA should never be a problem, at least in principle. Control amplifications on samples to show that a constant restriction enzyme site can be cut strengthen confidence in the analysis. Furthermore, the presence or absence of polymorphisms can be assessed immediately by agarose gel electrophoresis and by inspection of ethidium bromide stained products under ultraviolet light.

Uses of the reaction in molecular genetics

● Analysis of restriction fragment length polymorphisms
● Analysis of messenger RNA
● Amplification of fragments for identification by Southern blotting
● Assessment of genetic polymorphisms in linkage analysis
● DNA sequencing
● Preparation of elusive DNA fragments for cloning

The reaction has in fact permitted dramatic advances in the use of genetic polymorphism in linkage analysis. As well as classic dimorphic RFLPs, which are not particularly informative, several classes of repeat sequences have now been characterised in DNA which are highly polymorphic as a consequence of wide variation in repeat copy number[11] (see also chapter 16). This is easily assessed by the polymerase chain reaction using sequences that flank the microsatellite or minisatellite repeat—the size of the product reflects the number of repeat units. The procedure has revolutionised human genetics by massively increasing the speed and power of pedigree analysis. Of importance to the clinician, certain inherited diseases (myotonic dystrophy, the fragile X syndrome, Kennedy's disease) seem to be the result of spontaneous increases in the copy number of trinucleotide microsatellite repeats[12] (see also chapter 16). Again, assays for this based on the reaction already permit presymptomatic or antenatal diagnosis.

DNA sequencing

DNA sequencing, another fundamental molecular genetic technique and increasingly a vital component for identifying molecular pathology in patients, has also become reliant on the reaction. Traditionally, single stranded DNA for sequencing was obtained by cloning into the M13 bacteriophage system and subsequently purifying phage DNA. Given a reasonable amount of DNA template, sequencing was achieved by extending oligonucleotide primers in the presence of dideoxynucleoside triphosphate chain terminators. Amplification by the reaction solves the DNA yield problem without cloning. The material may be either sequenced double stranded or rendered single stranded by asymmetrical polymerase chain reaction (using very much less of one of the two primers),[13] selective enzymatic digestion of one of the strands, or by selective capture of one of the strands by labelling one oligonucleotide with biotin and immobilising the product on streptavidin coated magnetic beads. So called "cycle sequencing" can increase the sensitivity of the process. In many circumstances the use of high temperatures for the thermostable DNA polymerase reaction is advantageous in sequencing because secondary structure in the template strand is eliminated.

Contamination

There can be problems in controlling this powerful technique. As the polymerase chain reaction is able to amplify even a single molecule, contamination by any previously amplified DNA would be catastrophic. Many precautions have been devised to avoid contamination,[14] but scrupulous housekeeping is essential in any laboratory routinely undertaking the reaction. An extensive debate about the fidelity of copying DNA sequences by *Taq* polymerase has been published, but this is not usually an issue except in highly specialised applications, such as the study of variant clinical isolates of HIV.[15] In such cases an alternative polymerase, such as PFu polymerase, which has a proofreading function, can be used to increase the accuracy of the PCR copy.

Some applications of the reaction

There is an enormous range of novel applications of the reaction, and the list of references contains several comprehensive reviews. The following are some specific examples that have been particularly useful.

Quantitative polymerase chain reaction, which permits measurement of the level of specific mRNAs in different cell populations.[16 17] The amount of target product of the reaction is compared with the amount generated from a known amount of control amplification target in the same reaction. Similarly, changes in the amount of an mRNA in a particular situation can be quantified by amplifying it from different samples and similar amplifications of a control mRNA whose level is known not to change in response to the stimulus. The technique has been most widely applied in the analysis of cytokine responses.

RACE polymerase chain reaction. RACE stands for rapid amplification of cDNA ends. The reaction is performed between one primer designed on the basis of a given protein sequence and a second, which flanks the cloning site in a phage, or plasmid vector. The substrate is a total cDNA library. The product of the reaction is derived from the target mRNA, and the technique is particularly useful for isolating full length cDNA clones and identifying 5′ terminal cDNA sequences.

Alu polymerase chain reaction is enormously valuable.[18] Reaction primers match sequences in the ubiquitous interspersed repetitive sequences, of which about a million copies are scattered throughout the human genome. Amplification yields reaction products derived from the regions between Alu sequences. As the Alu repeats are relatively unique to humans, the human component from sources such as somatic cell hybrids in rodent cell backgrounds, or human DNA cloned in cosmids or yeast artificial chromosomes (YACs), can be isolated.

Inverse polymerase chain reaction permits amplification of DNA in which the sequence of only one end of the fragment is known. This permits effective chromosome walking.[19 20]

Vectorette polymerase chain reaction also permits amplification of regions of DNA of unknown sequence that flank known sequences. The approach entails restriction digestion of the DNA sample then ligation of specially designed "vectorette" linkers. The amplification is performed between the known sequence and the vectorette sequence. This method has proved generally useful in chromosome walking exercises and is particularly convenient for isolating the ends of yeast artificial chromosome clones for rapid physical genetic mapping and ordering of overlapping YACs to assemble so called "contigs".[21]

Specific applications of the reaction in molecular medicine

There are three distinct types of clinical challenge for which the polymerase chain reaction is indispensable:

1 Detecting vanishingly small amounts of DNA so as not to miss even the most cryptic infection and permit analysis of single cells or single sperm and of partially degraded samples.
2 Identifying the specific new mutation in a particular gene that causes a given inherited disease in a patient. This is necessary to gain an understanding of the molecular basis of the resulting disease and also to allow accurate family studies for genetic counselling.
3 Analysis to detect known mutations that always cause a particular inherited disease or polymorphism (for example, sickle cell anaemia or the cystic fibrosis F508 deletion).

Present uses of the reaction in medicine

- Detection of vanishingly small amounts of nucleic acid—for example, in HIV infection
- Identification of new genetic mutations
- Routine detection of known mutations—for example, Duchenne type muscular dystrophy, cystic fibrosis assays based on the reaction allow presymptomatic or antenatal diagnosis of diseases caused by microsatellite instability, such as myotonic dystrophy, the fragile X syndrome, and Kennedy's disease
- Detection of mutations thought to predispose to disease—for example, myocardial infarction and cancer
- Detection of mutations in malignant tumours to assess prognosis
- HLA subtyping

Detecting vanishingly small quantities of nucleic acid has been achieved in the diagnosis of HIV infection and in such unusual applications as detecting measles virus RNA in brain biopsy specimens from patients with subacute sclerosing panencephalitis. The nested technique is often helpful, and eventually the products of polymerase chain reactions may even be analysed by Southern blotting to gain further absolute sensitivity. In these circumstances, the issue of sample contamination giving false positive results is usually as important as the risk of false negative results arising from "failure" of the reaction. The ability to detect negligible (a few molecules) DNA is also essential for preimplantation diagnosis in in vitro fertilisation clinics.[22]

Characterisation of gene mutations

A range of elegant techniques has been developed for characterising new genetic mutations—all the techniques rely on the generation by the polymerase chain reaction of large quantities of DNA fragments from normal and mutant alleles.[23][24] The problem is to detect as little as a single base pair difference between, say, two 500 base pair fragments. This is often achieved by creating heteroduplexes between the two alleles and looking for modified properties in:

- chemical cleavage reactions
- ribonuclease digestion (one fragment is RNA)
- denaturing gradient gel electrophoresis
- single strand conformational polymorphism
- reaction with carbodiimide
- automated total sequencing of both fragments.

A smaller number of techniques based on the polymerase chain reaction are available for the routine detection of known mutations. If the mutation creates or destroys a restriction site then this can be simply examined in the products of the reaction. The disadvantage is that two distinct operations (polymerase chain reaction and restriction cleavage) are required. Detecting mutations by hybridising products of the reaction with allele specific oligonucleotides, containing either the wild type or mutant sequence, is possible, although again the need for two distinct procedures is a drawback.

The concept of an allele specific polymerase chain reaction, usually called "ARMS" (the amplification refractory mutation system), is rather more convenient and general.[25] Here one of the pair of reaction primers is deliberately designed so that its 3′ terminal residue lies precisely at the point mutation site. Two experiments are run in parallel: in one the allele specific primer matches the normal sequence; in the other it matches the mutant sequence. A product is obtained from the reaction only when 3′ residues form base pairs and prime synthesis correctly. By comparing the products in the "normal" and "mutant" reactions, normal and mutant homozygotes as well as heterozygote carriers are easily identified. Many genetic diseases are now routinely tested for in this way.

Furthermore, several pairs of the reaction primers can be mixed in the same tube and to "analyse" different point mutations at the same time (all the different products of the polymerase chain reaction are deliberately designed to be different sizes). By this "multiplex" approach, for example, all of the four (or more) common mutations causing cystic fibrosis can be looked for simultaneously in a single sample.[26] Several similar assays based on thermostable DNA ligases rather than polymerase are also being developed. A similar type of multiplex polymerase chain reaction analysis aimed at detecting DNA deletions has revolutionised antenatal diagnosis of Duchenne type muscular dystrophy. Another clinical advantage of these exquisitely sensitive tests has been that sufficient DNA can be obtained from buccal washings, dried blood on Guthrie phenylketonuria test cards, or tiny chorionic villous biopsy specimens.

Applications

As the technique has developed so the range of applications in clinical practice has expanded. HLA molecular subtyping by polymerase chain reaction is straightforward, and clinicians are now attempting to develop tests that predict people at risk of developing insulin dependent diabetes mellitus on that basis.[27] Mutations that appear to predispose to myocardial infarction (in the angiotensin converting enzyme (ACE) gene)[28] and hypertension (in the angiotensinogen gene)[29] have been identified recently. Population screening programmes may be worth while (see chapter 16). Both predispositions may be suppressible by treatment with ACE inhibitors. Various mutations in both oncogenes and tumour suppressor genes have been implicated heavily in the development of cancer. Assessment of the extent of such mutation in a given tumour may permit assessment of prognosis or predict whether distant metastasis has already occurred.

Future clinical possibilities

This raft of techniques can be applied to most specialties. Here I will limit discussion to two disciplines in which there is great activity.

Antenatal diagnosis

In antenatal diagnosis much effort is directed towards methods of analysing those few fetal cells found circulating in a pregnant woman's peripheral blood. By this approach invasive fetal sampling techniques may be avoided entirely or limited to those cases where confirmation of a positive diagnosis is desirable.

The estimate is that less than 10 000 fetal nucleated erythrocytes are present in 20 ml of maternal blood. Fluorescence activated cell sorting with anti-CD71, anti-CD36, or anti-glycophorin A, or a combination of these three monoclonal antibodies, permits enrichment towards 90% pure fetal nucleated erythrocytes. Trisomies can be detected direct by fluorescence in situ hybridisation. It seems to be only a matter of time before sufficient purification is achieved routinely to permit the full range of diagnosis based on the polymerase chain reaction as above. Indeed, this whole topic was resurrected by the demonstrations of Y chromosome specific product obtained by the technique from pregnant women with male fetuses.[30] This essentially confirmed that fetal tissue was in fact present in the maternal circulation.

Detection of cancer

Early detection of cancers is widely regarded as clinically important in that early treatment usually improves prognosis, and curative surgery may even be possible—for example, in colorectal cancer. The technique will have a major impact in this subject. It is already clear that mutations in p53 genes indicative of developing bladder cancer can be detected by analysis of shed cells present in patients' urine samples.[31] Even more remarkably, premalignant changes in the gastrointestinal tract can be detected by testing faeces using the polymerase chain reaction. Enough premalignant cells are present in the bulk of stool to permit the analysis of tumour suppressor gene mutations by this technique. Patients at high risk can thus be selected for colonoscopy. Whether this approach offers advantages over faecal occult blood testing remains to be established.[32 33] The fact that gastric *Helicobacter pylori* infection has been assessed from stool samples by this technique suggests that gastric malignancy may also eventually prove detectable in this way. Detection of potentially metastatic cells in the circulation of patients with newly diagnosed primary tumours (for example, melanoma) is particularly intriguing.[34] The characterisation of a melanoma related tumour suppressor gene from human chromosome 9 should make this type of approach even more useful.[35] The clinical usefulness of the polymerase chain reaction thus seems to be limited only by the power of our imagination in identifying specific targets.

1 White TJ, Arnheim N, Erlich HA. The polymerase chain reaction. *Trends Genetics* 1989;5:185–9.
2 Innis MA, Gelfand DH, Sninsky JJ, White TJ, eds. *PCR protocols*. New York: Academic Press, 1990.
3 McPherson MJ, Quirke P, Taylor GR, eds. *PCR, a practical approach*. Oxford: Oxford University Press, 1991.
4 Latchman DS, ed. *PCR applications in pathology*. Oxford: Oxford University Press, 1995.
5 Saiki RK, Scharf S, Faloona F, Mullis KB, Horn GT, Erlich HA. Enzymatic amplification of β-globin genomic sequences and restriction site analysis for diagnosis of sickle cell anemia. *Science* 1985;230:1350–4.
6 Saiki RK, Gelfand DH, Soffel S, Scharf SJ, Higuchi R, Horn GT, *et al*. Primer-directed enzymatic amplification of DNA with a thermostable DNA polymerase. *Science* 1988;239:487–9.
7 Kleppe K, Ohtsuka E, Kleppe R, Molineux L, Khorana HG. Studies on polynucleotides XCVI: repair replications of short synthetic DNAs as catalysed by DNA polymerases. *J Mol Biol* 1971;56:341.
8 Chien A, Edgar DB, Trela JM. Deoxyribonucleic acid polymerase from the extreme thermophile Thermus aquaticus. *J Bacteriol* 1976;127:1550.
9 Frohman MA, Dush MK, Martin GR. Rapid production of full-length cDNAs from rare transcripts: amplification using a single gene-specific oligonucleotide primer. *Proc Natl Acid Sci USA* 1988;85:8998–9002.
10 Loh EY, Elliott JF, Cwirla S, Lanier LL, Davis MM. Polymerase chain reaction with single-sided specificity: analysis of T cell receptor delta chain. *Science* 1989;243:217–20.
11 Weber JL, May PE. Abundant class of human DNA polymorphisms which can be typed using the polymerase chain reaction. *Am J Hum Genet* 1989;44:388–96.
12 Richards RI, Sutherland GR. Dynamic mutations: a new class of mutations causing human disease. *Cell* 1992;70:709–12.
13 Gyllesten UB, Erlich HA. Generation of single-stranded DNA by the polymerase chain reaction and its application to direct sequencing of the HLA-DQA locus. *Proc Natl Acad Sci USA* 1988;85:7652–6.
14 Sarkar G, Sommer SS. Shedding light on PCR contamination. *Nature* 1990;343:27.
15 Tindall KR, Kunkel TA. Fidelity of DNA synthesis by the *Thermos aquaticus* DNA polymerase. *Biochemistry* 1988;27:6008–13.
16 Chelly J, Kaplan JC, Maire P, Gautron S, Kahn A. Transcription of the dystrophin gene in human muscle and non-muscle tissue. *Nature* 1988;333:858.
17 Siebert PD, Larrick JW. Competitive PCR. *Nature* 1992;359:557–8.
18 Nelson DL, Ledbetter SA, Corbo L, Victoria MF, Ramirez-Solis R, Webster TD, *et al*. Alu polymerase chain reaction: a method for rapid isolation of human-specific sequences from complex DNA sources. *Proc Natl Acad Sci USA* 1989;86:6686–9.
19 Triglia T, Peterson MG, Kemp DJ. A procedure for in vitro amplification of DNA segments that lie outside the boundaries of known sequences. *Nucl Acids Res* 1988;16:8186.

20 Ochman H, Gerber AS, Hartl DL. Genetic applications of an inverse polymerase chain reaction. *Genetics* 1988;**120**:621–3.

21 Riley JH, Butler R, Ogilvie D, Finniear R, Jenner DE, Powell S, *et al.* A novel rapid method for the isolation of terminal sequences from yeast artificial chromosome (YAC) clones. *Nucl Acids Res* 1990;**18**:2887–90.

22 Handyside AH, Pattinson JK, Penketh RJA, Delhanty JDA, Winston RML, Tuddenham EGD. Biopsy of human preimplantation embryos and sexing by DNA amplification. *Lancet* 1989;**i**:347–9.

23 Cariello NF, Skopek TR. Mutational analysis using denaturing gel electrophoresis and PCR. *Mutat Res* 1993;**288**:103–12.

24 Cotton RGH. Current methods of mutation detection. *Mutat Res* 1993;**285**: 125–44.

25 Newton CR, Graham A, Heptinstall LE, Powell SJ, Summers C, Kalsheker N, *et al.* Analysis of any point mutation in DNA. The amplification refractory mutation system (ARMS). *Nucl Acids Res* 1989;**17**:2503–16.

26 Ferrie RM, Schwarz MJ, Robertson NH, Vandin S, Super M, Malone G, *et al.* Development, multiplexing and application of ARMS tests for common mutations in the CFTR gene. *Am J Hum Genet* 1992;**51**:251–62.

27 Todd JA, Bell JI, McDevitt HO. HLA-DQ beta gene contributes to susceptibility and resistance to insulin-dependent diabetes mellitus. *Nature* 1987;**329**: 599–604.

28 Cambien F, Poirier O, Lecerf L, Evans A, Cambou JP, Arveiler D, *et al.* Deletion polymorphism in the gene for angiotensin-converting enzyme is a potent risk factor for myocardial infarction. *Nature* 1992;**359**:641–4.

29 Jeunemaitre X, Soubrier F, Kotelevtsev YV, Lifton RP, Williams CS, Charru A, *et al.* Molecular basis of human hypertension: role of angiotensinogen. *Cell* 1992;**71**:169–80.

30 Lo Y-MD, Patel P, Wainscoat JS, Sampietro M, Gillmer MD, Fleming KA. Prenatal sex determination by DNA amplification from maternal peripheral blood. *Lancet* 1989;**ii**:1363–5.

31 Sidransky D, Von Eschenbach A, Tsai YC, Jones P, Summerhays I, Marshall F, *et al.* Identification of p53 gene mutations in bladder cancers and urine samples. *Science* 1991;**252**:706–9.

32 Powell SM, Zilz N, Beazer-Barclay Y, Bryan TM, Hamilton SR, Thibodeau SN, *et al.* APC mutations occur early during colorectal tumorigenesis. *Nature* 1992;**359**:235–7.

33 Sidransky D, Tokino T, Hamilton SR, Kinzler K, Levin B, Frost P, *et al.* Identification of *ras* oncogene mutations in the stool of patients with curable colorectal tumors. *Science* 1992;**256**:102–5.

34 Smith B, Selby P, Southgate J, Pittman K, Bradley C, Blair GE. Detection of melanoma cells in peripheral blood by means of reverse transcriptase and polymerase chain reaction. *Lancet* 1991;**ii**:1227–9.

35 Fountain JW, Graw SL, Kao W, Stanton VP, Aburatani H, Munroe DJ, *et al.* Further characterization of the 9p21 region frequently deleted in human cutaneous melanoma. *Am J Hum Genet* 1992;**51**:A51.

4: An introduction to cells

L Wolpert

Cells are the triumph of evolution. The rest of evolution can be thought of as an elaboration on this masterpiece. The original cells were prokaryotes like the modern bacteria, and lack a nucleus and other organelles such as mitochondria. All animals are constructed from eukaryotic cells which have a well defined nucleus, chromosomes that condense at mitosis, and organelles such as mitochondria, endoplasmic reticulum, Golgi apparatus, centrioles, centrosomes, and a cytoskeleton. In some ways, cells are more complex than the organs to which they give rise, with the possible exception of the brain, in that their behaviour reflects the integrated activity of thousands of genes, their products, and the complex biochemical and structural networks that result.

In this biochemical network there are two different time scales. The first, which responds to changes in seconds or fractions of a second, is that concerned with metabolism. The enzymes in the cell cytoplasm, including the mitochondria, catalyse molecules along narrowly defined reaction pathways, such as the citrate cycle, the synthesis of purines, or the breakdown of carbohydrates. Many of these reactions require or generate energy in the form of ATP. The speed of response of these metabolic pathways can be contrasted with those in the second system, which entails the synthesis of macromolecules such as nucleic acids and proteins. Here the response times are minutes to hours. Understanding the integration of these two interdependent pathways is a major problem in cell biology.

Proteins characterise cells

The human body contains at least 200 different cell types. The character of a cell is determined by the proteins it contains. For example, the special feature of parenchymal liver cells is that they synthesise albumin, whereas lymphocytes synthesise immunoglobulins. The enzymes in a cell determine its metabolic pathways. The control of protein synthesis is thus central to the life of the cells and is controlled by nuclear-cytoplasmic interactions. It should be noted that cell biochemistry is largely based on protein complexes, rather than single proteins acting individually.

The genes on the chromosomes in the cell nucleus dominate the life of the cell by controlling protein synthesis, but they are curiously passive. The DNA of the genes codes only for proteins or for RNA itself. Typically, a DNA sequence is transcribed into a messenger RNA (mRNA) sequence, which is then processed within the nucleus, non-coding sequences known as introns being removed. It is this processed RNA that enters the cytoplasm as mRNA, where it is translated into the amino acid sequence of the protein on ribosomes. Control of protein synthesis is largely at the level of transcription itself, although in some cases control involves processing of the RNA and even control of translation of the message. The genetic control of cell activity is entirely the result of the control of protein synthesis, and thus the genes do not exert an immediate effect on the life of the cell. If the nucleus is removed from the cell, then, for example, the metabolic pathways are unaffected until there is a change in enzyme concentration. This will depend on the stability of both the proteins and their mRNA.

DNA and the cell

As there is good evidence that the DNA is the same in almost all cells in the body of an individual, there must be mechanisms that turn genes on and off in specific cells. Embryonic development can thus be viewed in terms of the control of differential gene activities, DNA providing a genetic programme. As development proceeds different cell types with different protein compositions emerge. Not only must genes be turned on and off, but this activity must also be stable so that it can be inherited when the cells multiply. Liver cells divide to give more liver cells.

Changes in DNA can alter cell behaviour. Mutations alter the sequence of the bases. This is what happens in sickle cell anaemia, in which the haemoglobin molecules have abnormal properties. A change in one of the codons of the gene for haemoglobin alters one amino acid, glutamine, to valine. This in turn changes the way the molecule folds and results in the haemoglobin molecules sticking together and forming rod like structures, which deform the red cell, giving it its sickle shape. This alters the flexibility of the cell and so affects its passage through fine capillaries, which results in anaemia. Thus, from a change in just one base there is a cascade of events leading to a physiologically abnormal condition.

Sickle cell anaemia is an excellent example of the relationship between DNA and function and form. In contrast to the intimate relationship between DNA and protein, the pathways whereby a protein affects the working of the cell and the body as a whole can be intricate, tortuous, and multifarious. In general these pathways are not known, and this is a major problem in cell biology. Even if we knew the complete sequence of the DNA of human cells we could not, at present, interpret it properly. Nevertheless the human genome project will enormously facilitate progress towards understanding of how DNA controls cellular functions. Already there are at least 4000 single mutations known to cause human malfunction.

Control of gene activity

A central problem is control of transcription of particular genes. This involves protein transcription factors which are synthesised in the cytoplasm and then move to the nucleus. Such proteins bind to control regions adjacent to particular structural genes and allow them to be transcribed in specific tissues. It is possible, for example, to join the growth hormone gene (normally active only in the pituitary) to the elastase control region (elastase is made only in the pancreas). When this construction is incorporated into the genome by injecting the DNA into the nucleus of a fertilised mouse egg, then growth hormone is also made in the pancreas of the transgenic mouse showing that there are special transcription factors in the pancreas controlling the transcription of the elastase gene.

For some cells, at least, the pattern of gene expression is reversible. Various different human cells, such as liver cells, can be fused with mouse muscle cells. When the liver nucleus is exposed to muscle cytoplasm then genes specific for muscle begin to be transcribed and human muscle proteins synthesised. It is clear that the pattern of gene expression remains sensitive to changes in transcription factors.

Protein targeting

The different proteins must be transported to the specific cellular compartments where they will be used: enzymes of the citrate cycle to the mitochondria and degradative enzymes to liposomes. This is achieved by the extensive membrane bound system in the cytoplasm, particularly the endoplasmic reticulum and the Golgi apparatus. To achieve this sorting, specific amino acid sequences serve as address markers. For example, if the amino acid terminal end carries the so called signal sequence, then the proteins are transferred across the membrane of the endoplasmic reticulum and can be secreted. Those that do not have the sequence remain in the cytosol. Further sorting then occurs so that, for example, some of those that enter the endoplasmic reticulum go through the Golgi apparatus to the plasma membrane or special secretory vesicles. It is clear that there is considerable organisation of the internal membrane system in the cytoplasm.

Cytoskeleton

A major change in our concept of the cell relates to the structure of the cytoplasm or, more fashionably, the cytoskeleton. A substantial fraction of the cellular proteins makes up a fibrous network, which seems to serve as a scaffold. This scaffold organises many other cell constituents and is concerned with both maintaining and altering the shape of the cell and with bringing about cell movement. Major components here are actin filaments, microtubules, and intermediate filaments. Intermediate filaments are stable compared with microtubules and actin filaments, which are continually assembling and disassembling. There is a large research industry devoted to analysing this process, particularly in relation to motility and the role of other proteins. There is also a major effort towards understanding how the cytoskeleton is linked to the plasma membrane. One view is that events at the surface may be transduced through the cytoskeleton to provide a signal to the cell nucleus. This is closely linked with the recent emphasis given to the possible importance of cell shape and the role of the extracellular matrix in determining cell behaviour.

Extracellular matrix and plasma membrane

Rather than being perceived as merely rather dull packing, the extracellular matrix, which is secreted by the cells, is now seen as having a major role in determining cellular behaviour. This is particularly true of epithelia. For example, mammary gland epithelium in culture behaves quite differently depending on the nature of the matrix to which it is attached. This interaction may be mediated by the matrix determining cell shape. Again, the migration of cells is particularly dependent on the matrix. Major matrix components include fibronectin and laminin, and there are at least ten different collagens. A major component of the matrix is the glycosaminoglycans.

The plasma membrane is now considered in terms of a fluid bimolecular lipid membrane in which proteins, such as receptors, float, move, and interact. These proteins provide the mechanism for the selective entry of molecules into the cell. Lipid soluble substances can diffuse across the membrane, but there are special transport systems for ions and molecules such as glucose and amino acids. Larger molecules can enter by endocytosis.

Of particular importance is signal transduction when signal molecules bind to receptors in the cell membranes, for example, growth factors bind to receptors that activate a phospholipase on the inner surface. This splits phospholipid into inositol triphosphate and diacylglycerol. Inositol triphosphate leads to an increase in calcium ions, whereas diacylglycerol activates protein kinase. The remainder of the pathway is complex and involves many further kinases and phosphatases.

With some exceptions, such as steroid hormones, in general we do not know the sequence of events that lead from an external signal to the cell's response. The sequence for glucocorticoid is well established: it enters the cell, binds a cytoplasmic protein, enters the nucleus, and attaches to specific sequences of DNA activating genes. But in many other cases of signal transduction, including the action of insulin, growth hormones, and adrenaline, for which some of the early steps are known to be complex, the sequence is only partly understood.

Conclusion

A major problem is to try and understand how all the cell's activities are integrated so that cells respond, multiply, differentiate, and move, in an appropriate manner. There are, for example, thousands of proteins involved in signal transduction. Cells are not what they used to be. Their study has been transformed by new techniques, particularly the analysis of cell components in molecular terms and the all persuasive influence of recombinant DNA technology. Remarkable progress has been made in isolating, breaking up into new components, and reassembling cellular structures and functions in vitro. Of particular importance are the new techniques for altering the cell's genetic constitution—both removing specific genes and adding new ones. Perhaps the assembly of whole and new kinds of cells is not too far off. Will our grandchildren have 48 chromosomes, the new ones carrying prophylactic genes?

Further reading

Alberts B, Bray D, Lewis J, Raff M, Roberts K, Watson JD. *The molecular biology of the cell*, 3rd edn. London: Garland, 1994.

5: Stem cells in normal growth and disease

B R Clark, T M Dexter

Each day, under normal conditions, the regenerating tissues of the adult human lose more than 100 g of mature cells.[1][2] To maintain the integrity of these tissues in response to this constant cell loss, the mature cells must be replaced. The regenerating tissues (for example, gastrointestinal epithelium, bone marrow, skin, testes) also have the capacity to produce extra cells in response to conditions of additional cell loss caused by bleeding, infection, or injury. Mature cells produced by these systems are, however, highly specialised cells and are usually incapable of further growth. The numbers of mature cells are maintained by the proliferation and development of more primitive cells, known as stem cells. The process by which stem cells develop into their clearly distinguishable mature cell counterparts is termed "differentiation". Regenerating tissues are maintained by the persistence of a stem cell population which gives rise to differentiated progeny under carefully regulated conditions. This chapter will examine the biology of regenerating tissues and highlight clinical conditions, such as cancer, where these processes are dysregulated and lead to abnormal cell growth.

Origin of stem cells

Two models have been proposed to explain the persistence of stem cells throughout life.[3-5] The first model suggests that a fixed number of primitive cells is laid down during embryogenesis to supply the body's needs throughout its lifetime. Stem cells are recruited into proliferation, differentiation, and development as required—like the recruitment of oocytes from the ovaries. In this model, as stem cells differentiate to contribute mature cells to the tissue, the stem cell pool declines. The second model also suggests that a small population of stem cells arises during embryonic development, but that these cells can reproduce themselves (undergo self renewal) to produce daughter cells, which retain the same proliferative and developmental potential as the original parental cells. In this model, therefore, the recruitment of stem cells into proliferation and development does not necessarily lead to a reduction in the number of stem cells. The weight of evidence strongly supports this second model.[67] The critical point here is that life and death processes must be balanced—exit from the stem cell population as a consequence of differentiation or death must be balanced by an input of cells into the stem cell population. How, then, are tissues organised to maintain not only this basal balance but also the ability to respond in stress conditions? Many lessons have been learned from the haemopoietic system, and this will be used as a primary example to discuss aspects of stem cell biology and how this knowledge can be applied for therapeutic advantage.

Stem cells: capable of sustained self renewal

Haemopoietic stem cells are formed during embryogenesis and colonise the bone marrow where they contribute to neonatal and adult haemopoiesis.[8] Within the bone marrow cavities the stem cells associate with the complex range of cells and extracellular matrix of the bone marrow stroma.[9] Extensive work has shown that the different kinds of blood cells—neutrophils, monocytes and macrophages, eosinophils, basophils, erythrocytes, megakaryocytes (platelets), osteoclasts, B lymphocytes, and T lymphocytes—are all derived from a common stem cell pool in the marrow.[10-13] In other words, a haemopoietic stem cell is multipotent—capable of giving rise to multiple haemopoietic lineages. In other tissues, the presence of such a spectacularly multipotent stem cell is not quite so obvious. In the testes, for example, only spermatogonia are generated by sperm stem cells, and in the skin the epithelial stem cells give rise to mature keratinocytes only. However, stem cells in all tissues share a common feature—the rate of stem cell self renewal must be balanced against differentiation and mature cell formation and loss. It is this ability of stem cells to persist by self renewal that is common to all stem cells.[14]

Organisation of regenerating tissues

How do the stem cells "know" when to divide and generate cells capable of giving rise to mature cells to maintain the complement of cells in a tissue? How do stem cells maintain a constant stem cell pool by self renewal? The answers lie in the organisation of tissues.

Intestinal epithelium covers the surface of the intestinal tract. Intestinal epithelial stem cells are present in the crypts of Lieberkühn—a set of glands at the bases of the intestinal villi. Epithelial cells are generated in the crypts and migrate on to the villi to replace the mature epithelial cells that are shed or sloughed off. When crypts are examined in cross section using techniques to label the DNA of dividing cells, it becomes clear that the stem cells in the crypt—localised in specific sites—give rise to more differentiated progeny, which have a limited ability to undergo self renewal (fig 5.1).[15] These daughter cells expand in number and differentiate while migrating out of the crypt and on to the villi. The mature epithelial cells do not divide further and form the epithelium covering the villi. Thus, in the intestine, the largest contribution to mature cell numbers occurs through the expansion of the population of differentiating cells with a limited ability to self renew—the progenitor cell population.

Progenitor cells are found in other systems, such as the skin or bone marrow, where cell numbers are amplified. In haemopoiesis, for example, 3×10^{11}—that is, 300 000 million—cells are required each day.[2] It only requires 10 divisions of a single progenitor cell and its daughter cells to give rise to over 1000 cells (2^{10}; see fig 5.2). It is the expansion of the progenitor population—and not the stem cells—that gives rise to the cellular amplification seen in several tissues.

Nevertheless, why are stem cell pools not depleted as they generate more progenitor cells throughout a lifetime? The most popular hypothesis to account for this suggests that there exists a finite number of sites in tissues where stem cells can maintain their primitive phenotype—because of a complex requirement for accessory cells and extracellular matrix.[5] In tissues such as the

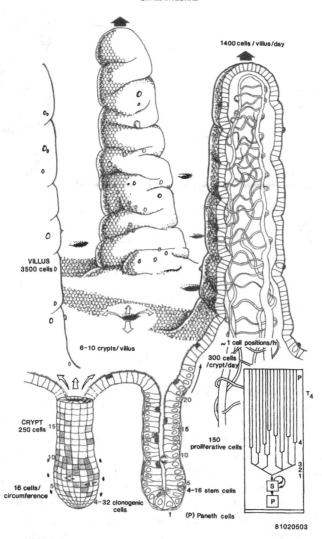

FIG 5.1—*The structure and maintenance of the small intestine. The boxed region illustrates the relationship between a self renewing stem cell (S) and endothelial cells that proliferate and migrate out of the crypt. In this example, four rounds of cell division generate 16 maturing endothelial cells. Paneth cells (P), which contribute to mucosal barrier function are also generated and migrate downwards from the stem cell zone. (Reprint by kind permission of Professor C S Potten.)*

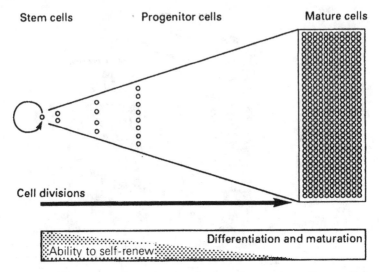

FIG 5.2—*Diagram to show the contribution of progenitor cells to cellular amplification. Progenitor cells and their progeny undergo division to give rise to the mature cell populations. This ability to self renew declines as the cells undergo differentiation into mature cells. For clarity, not all division steps are shown.*

gastrointestinal epithelium and skin, stem cells are localised to discrete regions of the tissue and can be shown to be present in a finite number, indicating that the sustained generation of mature cells requires stem cell self renewal.[15 16] For the purpose of this discussion, the term "niche" will be used to describe the finite, privileged environment that maintains any stem cell in a primitive state. Niches act to limit and maintain the size of the stem cell pool, ensuring that it is not doubled after stem cell self renewal. Only one of the daughter cells can remain in the privileged niche, where it preserves its capacity to survive and self renew. The other daughter cell moves out of the niche to differentiate or die.

Indeed, cell death is a common phenomenon in many tissues, where it may be used to regulate the numbers of stem cells, progenitor cells, and mature cells.[17 18] In this instance, cell death is not simply a result of the cell wasting away through lack of nutrition or anchorage. This regulatory cell death appears to be an active decision to die swiftly when an essential supportive stimulus is removed, or when the cell receives a signal instructing it to die.[18] This process of active cell death is termed "apoptosis". The swift destruction of a cell undergoing death by apoptosis—as opposed to death by wasting—is caused by the start of a "self destruct" sequence of events, which destroys the cell and packages the pieces in such a way that they can be scavenged efficiently.[19] Apoptosis ensures that cells with a capacity to self renew do so in a controlled manner—under the influence of the appropriate signals—and that cells in the wrong place or surviving at the wrong time are promptly and efficiently destroyed. Many examples of apoptosis are seen in tissue destruction and remodelling. Perhaps one of the most obvious is seen in the generation of fingers from fetal limb paddles, when interdigital cells are lost in a controlled manner to give a five fingered hand.[17] (See also chapter 7.)

Systemic and local regulation of haemopoiesis

The bone marrow is a complex three dimensional structure and its spatial organisation is only now being elucidated. The structural elements of the marrow—the stroma—form an extended web like structure within the marrow cavity. This composition has not facilitated analysis of the marrow environment in vivo. For the moment, it is unclear whether the marrow has discrete regions comparable to gut epithelium, which are thought to provide regulation to the components of the renewing tissue.

In the haemopoietic system, events such as bleeding or coagulation decrease the number of circulating mature cells and result in compensatory changes within the bone marrow which normalise the levels of circulating cells. The number of circulating cells can also be increased in response to, for example, infection or a decline in atmospheric pressure (for example, living at elevated altitudes). How is the haemopoietic system regulated to generate the required numbers and types of mature cells in response to such diverse stimuli? In addition to the largely undefined role of the marrow stroma, haemopoiesis can be influenced by a wide variety of systemic factors which act upon haemopoietic progenitor cells in the bone marrow.

Perhaps the most striking example of systemic regulators in haemopoiesis involves the production of mature red blood cells (RBCs). It has been known for nearly 100 years that the body produces a hormone, erythropoietin, which can stimulate the generation of red blood cells.[21 22] Erythropoietin is a glycoprotein hormone generated by kidney cells in response to changes in the local oxygen tension. Erythropoietin is essential for the development of red blood cells from erythroid progenitors in the marrow. In the absence of erythropoietin, erythroid progenitors are thought to die by apoptosis. Thus the supply of erythropoietin regulates the production of mature erythrocytes.

TABLE 5.1—*Some (by no means all) examples of growth factors, negative regulators, and differentiation inducing agents*

Growth factor	Some biological actions	Some clinical applications
G-CSF	Stimulation of neutrophil progenitors, neutrophil production, and function of mature neutrophils	Preventing neutropenia after chemotherapy, bone marrow transplantation, or in conditions of congenital or acquired neutropenia, for example, cyclic neutropenia, AIDS
GM-CSF	Stimulation of neutrophil progenitors, neutrophil production	As G-CSF
Erythropoietin	Stimulation of erythropoiesis	Treating anaemia of chronic renal disease
M-CSF	Stimulation of myelopoiesis	Enhancing recovery of myeloid cells after bone marrow transplantation
SCF	Acts on very primitive haemopoietic cells to enhance responses to other growth factors	Enhancing response to other growth factors
IL2	Stimulates T lymphocytes, influences capillary endothelium	Lymphokine activated killer (LAK) cell treatment of malignancies. Response seen with IL2 alone in renal cell cancer
IFN-α	Interferons act on cell division and differentiation. Discovered as compounds with in vitro antiviral activity	Decreasing Philadelphia chromosome positive cells in chronic myeloid leukaemia (CML), inducing remission in hairy cell leukaemia
IFN-γ		Some activity on haematological and solid malignancies
TGF-β	Decreases cell cycling in primitive progenitor cells in bone marrow and gut	Protecting gut and marrow stem cells during cycle specific chemotherapy
MIP-1α	Decreases cell cycling in primitive bone marrow cells	Protecting gut and marrow stem cells during cycle specific chemotherapy
LIF/DIA	Inhibits cycling of leukaemic cells in vitro. Maintains primitive phenotype of embryonal stem cells	Possibly in stimulating platelet production

(continued)

TABLE 5.1—*Growth factors, negative regulators, differentiation inducing agents* *(continued)*

Growth factor	Some biological actions	Some clinical applications
Retinoic acid	Influences differentiation of cell types possessing suitable receptor	Treating some acute promyelocytic leukaemias, oral leucoplakia, acne, and squamous cell carcinoma
EGF	Stimulates formation of skin	Treating partial thickness skin wounds including burns

G-CSF = granulocyte colony stimulating factor, GM-CSF = granulocyte–macrophage colony stimulating factor, M-CSF = macrophage colony stimulating factor, SCF = stem cell factor (or c-kit ligand, mast cell growth factor), IL = interleukin, IFN = interferon, TGF = transforming growth factor, MIP = macrophage inflammatory protein, LIF/DIA = leukaemic inhibitory factor/ differentiation inhibitory activity, EGF = epidermal growth factor.

In addition to erythropoietin many other haemopoietic growth factors (HGFs) have been discovered.[23] The genes encoding these glycoproteins have been isolated, permitting the production of sufficient pure haemopoietic growth factors for clinical applications (table 5.1). The populations of haemopoietic progenitor cells that these factors influence can be identified through the use of in vitro culture assays. Under appropriate culture conditions, individual haemopoietic progenitor cells divide and give rise to colonies of cells, which may include more differentiated haemopoietic cells. A cell that gives rise to a colony is termed a "colony forming cell" (CFC), and the type of cells found within the mature colony indicates the type of colony forming cell. Some of the colonies contain cells of multiple lineages—for example, erythroid cells, neutrophils, and megakaryocytes—and are obviously derived from a multipotent progenitor cell. Other colonies contain mature cells representative of only one or two cell lineages and are derived from unipotent or bipotent progenitor cells. For example, a colony containing granulocytes and macrophages is generated by a bipotent granulocyte–macrophage colony forming cell (GM-CFC). Some haemopoietic growth factors are termed "colony stimulating factors" (CSFs) because of their action upon colony forming cells in these in vitro assays. For example, GM-CSF supports the formation of colonies by progenitor cells capable of differentiation into granulocytes and macrophages.

Based on the in vitro colony forming data, colony forming cells can be placed into a scheme of haemopoiesis within which multipotent and bipotent progenitor cells become committed to a

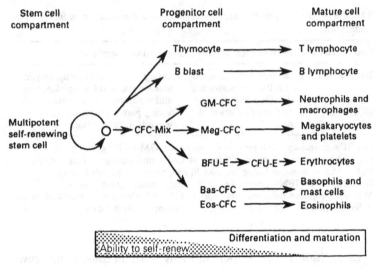

FIG 5.3—*Diagram showing the differentiation and development of stem cells in a haemopoietic hierarchy. (CFC-Mix = a multipotential cell producing mixed myeloid colonies in vitro, GM-CFC = granulocyte and macrophage colony forming cell, BFU-E = a primitive erythroid progenitor cell, CFU-E = a more mature erythroid colony forming cell, Eos-CFC = eosinophil colony forming cell, Bas-CFC = basophil and mast cell progenitor, Meg-CFC = megakaryocyte colony forming cell.)*

single lineage (fig 5.3). These lineage committed progenitors and their daughter cells have a limited ability to self renew, but they can divide to give rise to the cellular amplification seen in haemopoiesis.

Haemopoietic growth factors such as SCF and Epo can be found in the serum at levels which are known to stimulate colony formation in vitro. However, many haemopoietic growth factors are thought to be produced within the marrow by cells of the bone marrow stroma. Some haemopoietic growth factors synthesised by stromal cells are produced as transmembrane glycoproteins and are found on the stromal cell surface.[24 25] Other haemopoietic growth factors are produced as soluble glycoproteins but may become sequestered by components of the bone marrow extracellular matrix.[26 27] Both of these localised, non-diffusing sources of haemopoietic growth factors are available to the developing haemopoietic cells.[24 26 27] Thus the cells and extracellular matrix of the bone marrow form a complex microenvironment for haemopoietic cells. However, this local control of haemopoiesis can be overridden by haemopoietic growth factors which are produced in response to stress conditions.

For example, haemopoietic growth factors such as G-CSF and GM-CSF are produced at sites of an immune response to infection.[28] These haemopoietic growth factors act on the bone marrow to promote the generation and release of neutrophils. One additional feature of many haemopoietic growth factors is that they also act to stimulate the function of mature cells. For example, G-CSF and GM-CSF enhance neutrophil metabolism and adhesion.[29-32] Thus growth factors produced at sites of infection may enhance the function of neutrophils in that region as well as promoting the production of more neutrophils and their migration to inflammatory sites.

In examples examined so far, cells such as erythrocytes or gastrointestinal epithelium are produced in response to cell loss. Feedback mechanisms—such as the detection of anoxia or perhaps the availability of space to migrate—act on progenitor cells to regulate the production of mature cells. In response to infection or wounding, stimulatory factors ensure the rapid production and migration of additional cells. However, in haemopoiesis several populations of cells exist that are very small in number and are generally quiescent. Lymphocytes, for example, constitute around 0·05% of circulating cells. Thus, unlike erythrocytes, a shortage of the normally quiescent lymphocytes may not be detected immediately and may only become apparent by a secondary means—for example, infection.

How are the numbers of lymphocytes regulated? As in erythroid development, apoptosis has a role in basal lymphopoiesis. More than the daily requirement of developing T and B lymphocytes is generated. The numbers of both cell types are regulated by the lymphoid organs—the thymus and regions of the gut for T lymphocytes in the adult and the spleen and secondary lymphoid organs for B lymphocytes. Some 95–99% of all developing T lymphocytes produced in the thymus (thymocytes) are thought to receive a "death" signal and die by apoptosis.[33] Only thymocytes that recognise foreign antigens—and not self antigens—survive and proliferate. The T lymphocytes leaving the thymus survive for many months or years. On the other hand, B lymphocytes have a much shorter half life—weeks to months—and must be renewed constantly.[34] Newly produced, virgin B lymphocytes leaving the marrow are thought to have been selected already, at least in part, for the ability to respond to foreign antigens only. The challenge for a virgin B lymphocyte is to receive a "survival signal" in the spleen or a secondary lymphoid organ.[35] Failure to receive such a signal is thought to result in death by apoptosis. Virgin B lymphocytes compete with the recirculating B lymphocyte pool for these signals. If the mature B lymphocyte pool is depleted then more of the virgin B lymphocytes released from the marrow can receive a survival signal. Thus, by using competition at the level of the lymphoid organs, the recruitment of new mature B lymphocytes is directly linked to the size of the existing pool.

Stem cells in disease

Dysregulation of stem cells and their more differentiated progeny is thought to contribute to several clinical conditions. The balance between cell production and cell loss that is normally tightly regulated in tissues can be subverted at any point from the level of stem cell self renewal to the inappropriate survival of maturing cells (table 5.2).

TABLE 5.2—*Some examples of conditions (many premalignant) caused by dysregulated growth in a range of tissue types*

Condition	Comment
Myelodysplastic syndrome	Defective haemopoietic cell production by stem cell pool. Haemopoiesis commonly clonal—the product of a single stem cell. Premalignant condition, evidence for transformation to acute myeloid leukaemia
Secondary polycythaemia	Overproduction of erythrocytes due to excess levels of erythropoietin
Polyposis coli	Premalignant condition that predisposes to colorectal carcinoma
Warts	Dysregulated production of epithelial cells caused by papillomaviruses. Papillomaviruses and genital warts are linked to increased risk of cervical carcinoma
Lentigo maligna	Defect of melanocytogenesis predisposing to malignant melanoma
Some small cell lymphomas	The *bcl-2* gene which suppresses apoptosis in developing B lymphocytes is involved in chromosomal translocations. This translocation directs inappropriate expression of the *bcl-2* gene, resulting in B lymphocyte survival in the absence of a "survival signal" and lymphoid proliferation localised to the lymph node

In benign neoplastic conditions, many so called premalignant lesions exhibit some degree of escape from the normal mechanisms that govern tissue turnover. For example, warts or intestinal polyps are formed when cells divide under the influence of abnormal growth regulation. These premalignant events can pave the way for critical mutations to genes responsible for growth regulation, which results in cellular transformation into a malignant neoplasm (see chapter 14).

Several genes that promote cellular division have been identified. Mutations or inappropriate expression of these genes are thought to be able to stimulate cells to divide. Some of these genes are called transforming oncogenes because of their suspected involvement in promoting malignant events.[36] Another set of genes acts to restrict cell division. Some are termed "anti-oncogenes" or tumour suppressor genes,[37] and in several malignancies tumour suppressor genes are lost or mutated. In vitro experiments have shown that, in such cases, the addition of the appropriate tumour suppressor gene(s) to cells from tumours results in a loss of a malignant phenotype, which suggests that the loss of tumour suppressor function contributed to the malignant transformation. A combination of mutations of both transforming oncogene and tumour suppressor gene is thought to be involved in the progression of polyposis to gastrointestinal carcinoma.[38]

Many clinical conditions are the result of dysregulated cell production (table 5.2). As the technology is not (yet) available to alter genetic lesions within cells in vivo, present attempts to influence the growth of cells rely on our knowledge of how these systems can be modulated by external stimuli. As discussed, dividing cells can be influenced by the surrounding environment—the presence of soluble factors that are adjacent to the cell. A wide range of soluble factors is known to act on tissues to influence aspects of cellular division or differentiation.[23 39] Some of these factors, termed "growth factors", are increasingly used in the management of conditions in which cellular regulation is dysfunctional or the body is unable to produce sufficient mature cells.

The therapeutic use of growth factors

Several growth factors, especially the haemopoietic growth factors, are in regular clinical use (see table 5.1). These factors have roles in modifying the generation and function of mature cells. Haemopoietic growth factors are applied to stimulate the production of mature cells when endogenous growth factors are reduced (for example, erythropoietin in chronic renal disease[40]) or when the haemopoietic system has been compromised—after bone marrow transplantation or damage by chemotherapy or radiation.[41] By expanding the number of progenitor cells and supporting differentiation into mature cells, haemopoietic growth factors enable compromised bone marrow to produce more mature cells to prevent infection or anaemia. In addition to promoting the generation of cells, growth factors can induce cells to begin the process of cell division. The cells undergoing division may be targeted by chemotherapy, enabling drugs to "hit" quiescent malignant cells that would otherwise escape treatment.[42]

In contrast, factors that act to suppress cellular division—termed "negative growth regulators"—may also have clinical applications (see table 5.1).[43-46] For example, transforming growth factor β (TGF-β) can inhibit the cell division of primitive progenitor cells in the gastrointestinal epithelium and bone marrow. These populations of cells are often damaged by radiation or chemotherapy used in cancer treatment to kill dividing cells. Suppressing cell division in the gastrointestinal epithelial or bone marrow progenitors can possibly reduce side effects and overcome some of the dose limiting toxicity, which permits treatment with larger doses. This approach will only work if the malignant cell population does not respond to the negative regulator and continues to divide. Some leukaemic cells do not respond to negative regulators such as TGF-β or MIP-1α (macrophage inflammatory protein 1α). Thus, MIP-1α may suppress the division of normal haemopoietic progenitor cells whereas leukaemic cells will continue to divide. Giving chemotherapy that kills dividing cells allows the leukaemic cells to be selectively targeted and the normal cells protected.[47]

The abnormal production of cells may be rectified in the future by using agents that promote differentiation. Immature cells that are undergoing self renewal may be induced to differentiate, giving rise to mature cells incapable of further division. This principle is currently in use against some dysplastic cutaneous lesions in which retinoic acid can act as a differentiating agent.[48 49] This general approach may be applicable to other diseases, such as acute myeloid leukaemias. In acute promyelocytic leukaemias, a translocation of chromosomes 15 and 17 places the gene for the retinoic acid receptor next to a region that directs its expression.[50] These leukaemias are very responsive to treatment which includes retinoic acid.[51]

The specificity of action of many growth factors makes them suited to tasks involving the modification of abnormal cellular growth and function. Future approaches to modifying cell growth and differentiation may be found in new techniques such as targeted gene treatment (see Chapter 20). Such an approach might allow the correction of many genetic defects by modifying tissues at the level of the stem cell.

1 International Commission on Radiobiological Protection. Report on the Task Group on Reference Man. Oxford: Pergamon, 1975: 144. (ICRP publication No 23.)
2 Cronkite EP, Feinendegen LE. Notions about human stem cells. Blood Cells 1976;2:263–84.
3 Hellman S, Botnick LE. Stem cell depletion: an explanation of the late effects of cytotoxins. Int J Radial Biol 1977;2:181–4.
4 Rosendaal M, Hodgson GS, Bradley TR. Organisation of haemopoietic stem cells: the generation-age hypothesis. Cell Tissue Kinet 1979;12:17–30.
5 Schofield R. The relationship between the haemopoietic stem cell and the spleen colony forming cell: a hypothesis. Blood Cells 1978;4:7–25.
6 Ross EAM, Anderson N, Micklem HS. Serial depletion and regeneration of the murine hematopoietic system: implications for hematopoietic organisation and the study of aging. J Exp Med 1982;155:432–44.
7 Harrison DE, Astle CM. Loss of stem cell repopulating ability upon transplantation. J Exp Med 1982;156:1767–79.
8 Metcalf D, Moore MAS. Haemopoietic cells. Amsterdam: North Holland Publishing Co, 1971.
9 Allen TD, Dexter TM, Simmons PJ. Marrow biology and stem cells. In: Dexter TM, Testa NG, Garland J, eds. Colony stimulating factors. New York: Marcel Dekker, 1990: 1–38.
10 Abramson S, Miller RG, Phillips RA. The identification in adult bone marrow of pluripotent and restricted stem cells of the myeloid and lymphoid lineages. J Exp Med 1979;145:1567–79.
11 Snodgrass R, Keller G. Clonal fluctuation within the haematopoietic system of mice reconstituted with retrovirus infected stem cells. EMBO J 1987;6:3955–60.
12 Till JE, McCulloch EA. Haemopoietic stem cell differentiation. Biochim Biophys Acta 1980;605:431–59.
13 Lemischka IR, Raulet DH, Mulligan RC. Developmental potential and dynamic behaviour of haemopoietic stem cells. Cell 1986;45:917–27.
14 Lajtha LG. Stem cell concepts. In: Potten CS, ed. Stem cells. London: Churchill Livingstone, 1983:1–11.
15 Potten CS, Hendry JH. Stem cells in the small intestine. In: Potten CS, ed. Stem cells. London: Churchill Livingstone, 1983:155–99.
16 Potten CS. Stem cells in epidermis from the back of the mouse. In: Potten CS, ed. Stem cells. London: Churchill Livingstone, 1983:200–32.
17 Duvall E, Wylie AH. Death and the cell. Immunol Today 1986;7:115–19.
18 Williams GT. Programmed cell death: apoptosis and oncogenesis. Cell 1991; 65:1097–8.
19 Savill J, Dransfield I, Hogg N, Haslett C. Vitronectin receptor-mediated phagocytosis of cells undergoing apoptosis. Nature 1990;343:170–3.

20 Lord BI. Cellular and architectural factors influencing the proliferation of hematopoietic stem cells. In: *Differentiation of normal and neoplastic hematopoietic cells*. Cold Spring Harbor, NY: Cold Spring Harbor Laboratory, 1978: 775–88.

21 Carnot P. Sur le mechanisme d'hyperglobulie provoquée par le serum d'animaux en renovation sanguine. *C R Acad Sci [III]* 1906;111:344–6.

22 Krumdiek N. Erythropoietic substance in the serum of anemic animals. *Proc Soc Exp Biol Med* 1943;54:14–17.

23 Metcalf D. The molecular control of cell division, differentiation commitment and maturation in haemopoietic cells. *Nature* 1989;339:27–30.

24 Anderson DM, Lyman SD, Baird A, Wignall JM, Eisman J, Rauch C, et al. Molecular cloning of mast cell growth factor, a hematopoietin that is active in both membrane and soluble forms. *Cell* 1990;63:235–43.

25 Rettenmier CW, Rousse MF, Ashmun RA, Ralph P, Price K, Sherr CJ. Synthesis of membrane-bound colony-stimulating factor 1 (CSF-1) and down modulation of CSF-1 receptors in NIH3T3 cells transformed by cotransfection of human CSF-1 and c-fms (CSF-1 receptor). *Mol Cell Biol* 1987;7:2378–87.

26 Gordon MY, Riley GP, Watt SM, Greaves MF. Compartmentalisation of a hemopoietic growth factor (GM-CSF) by glycosoaminoglycans in the bone marrow micro-environment. *Nature* 1987;326:403–5.

27 Roberts RA, Gallagher JT, Spooncer E, Allen TD, Bloomfield F, Dexter TM. Heparan sulphate-bound growth factors: a mechanism for stromal cell-mediated haemopoiesis. *Nature* 1988;332:376–8.

28 Metcalf D. *The molecular control of blood cells*. Cambridge, MA: Harvard University Press, 1988.

29 Yuo A, Kitagawa S, Ohsaka A, Saito M, Takaku F. Stimulation and priming of human neutrophils by granulocyte-colony stimulating factor and granulocyte-macrophage colony stimulating factor: qualitative and quantitative differences. *Biochem Biophys Res Commun* 1990;171:491–7.

30 Balazovich KJ, Almeida HI, Boxer LA, Recombinant human G-CSF and GM-CSF prime human neutrophils for superoxide production through different signal transduction mechanisms. *J Lab Clin Med* 1991;118:576–84.

31 Aranout MA, Wang EA, Clark SC, Sieff CA. Human recombinant granulocyte-macrophage colony-stimulating factor increases cell-to-cell adhesion and surface expression of adhesion promoting surface glycoproteins on mature granulocytes. *J Clin Invest* 1986;78:597–601.

32 Weisbart RH, Kwan L, Golde DW, Gasson JC. Human GM-CSF primes neutrophils for enhanced oxidative metabolism in response to the major physiological chemoattractants. *Blood* 1987;69:18–21.

33 Goldstein P, Ojcius DM, Young JD. Cell death and the immune system. *Immunol Rev* 1991;121:29–65.

34 Gray D, Skarvall H. B Cell memory is short lived in the absence of antigen. *Nature* 1988;336:70–3.

35 Cyster JG, Hartley SB, Goodnow CC. Competition for follicular niches excludes self reacting cells from the recirculating B-cell repertoire. *Nature* 1994;371: 389–95.

36 Bodmer W. Somatic cell genetics and cancer. *Cancer Surv* 1988;7:239–50.

37 Baker SJ, Markowitz S, Fearon ER, Wilson JK, Vogelstein B. Suppression of human colorectal carcinoma cell growth by wild type p53. *Science* 1990;249: 912–15.

38 Kinzler KW, Nilbert MC, Su-LK, Vogelstein B, Bryan TM, Levy DB, et al. Identification of FAP locus genes from chromosome 5q21. *Science* 1991;253: 661–5.

39 Cross M, Dexter TM. Growth factors in development, transformation and tumourigenesis. *Cell* 1991;6:271–80.

40 Sundal E, Businger J, Kappler A. Treatment of transfusion-dependent anaemia of chronic renal failure with recombinant erythropoietin. A European multicentre

study in 142 patients to define dose regimen and safely profile. *Nephrol Dial Transplant* 1991;**6**:955–65.

41 Sheridan WP, Lorstyn G, Wolf M, Dodds A, Lusk J, Maher D, *et al*. Granulocyte colony-stimulating factors and neutrophil recovery after high dose chemotherapy and autologous bone marrow transplantation. *Lancet* 1989;**ii**:891–5.

42 De Witte T, Muus P, Haanen C, Van der Lely N, Koekman E, Van der Locht A, *et al*. GM-CSF enhances sensitivity of leukemic clonogenic cells to long term low dose cytosine arabinoside with sparing of normal clonogenic cell. *Behring Inst Mitt* 1988;**83**:301–7.

43 Lord BI, Dexter TM, Clements JM, Hunter MA, Gearing AJH. Macrophage inflammatory protein protects multipotent haemopoietic cells from the cytotoxic effects of hydroxyurea in vivo. *Blood* 1992;**79**:59–63.

44 Keller JR, Ellingsworth LR, McNiece IK, Quesenberry PJ, Sing GK, Ruscetti FW. Transforming growth factor β directly regulates primitive murine hematopoietic cell proliferation. *Blood* 1990;**75**:596–602.

45 Keller JR, Sing GK, Ellingsworth LR, Ruscetti FW. Transforming growth factor β: possible roles in the regulation of normal and leukemic cell growth. *J Cell Biochem* 1989;**39**:175–84.

46 Ruscetti FW, Dubois C, Falk LA, Jacobsen SE, Sing G, Longo DL, *et al*. In vivo and in vitro effects of TGF-β1 on normal and neoplastic haemopoiesis. In: Bock GR, Marsh J, eds. *Clinical applications of TGF-β*. Chichester: Wiley, 1991:212–31. (Ciba Foundation Symposium No 157.)

47 Tsyrlova IG, Lord BI. Inhibitor of CFU-S proliferation preserves normal haemopoiesis from cytotoxic drug in long-term bone marrow culture—L1210 leukaemia model. *Leuk Res* 1989;**13**(suppl 1): 40.

48 Lippman SM, Parkinson DR, Itri LM, Weber RS, Schantz SP, Ota DM, *et al*. 13-Cis-retinoic acid and interferon alpha 2a: effective combination therapy for advanced squamous cell carcinoma of the skin. *J Natl Cancer Inst* 1992;**84**: 235–41.

49 Edwards L, Jaffe P. The effect of topical tretinon on dysplastic nevi. A preliminary trial. *Arch Dermatol* 1990;**126**:494–9.

50 Kakizuka A, Miller WH, Umesono K, Warrell RP, Frankel SR, Murty VV, *et al*. Chromosomal translocation t(15;17) in human acute promyelocytic leukemia fuses RAR alpha with a novel putative transcription factor, PML. *Cell* 1991;**66**: 663–74.

51 Castaigne S, Chomienne C, Daniel MT, Ballerini P, Berger R, Fenaux P, *et al*. All-trans retinoic acid as a differentiation therapy for acute promyelocytic leukemia. I. Clinical results. *Blood* 1990;**76**:1704–9.

6: Cell reproduction

Paul Nurse

The sequence of events and processes leading to cell reproduction is known as the cell cycle. During the cell cycle there is generally a duplication of cellular components followed by their partitioning into two daughter cells at cell division. For many components present in moderate to large amounts within the cell, this duplication and partition need not be controlled exactly. The chromosomes must, however, be precisely replicated and segregated during each cell cycle to produce two viable daughter cells. Replication of the DNA making up the chromosomes occurs during the S phase, and segregation of the chromosomes occurs during mitosis. These are the two major events of all cell cycles. In most cell cycles there are also gaps between these events: a G1 gap after mitosis and before the S phase and a G2 gap after the S phase and before mitosis.

In this chapter, there is a description of what is known about the controls that regulate the onset of S phase and mitosis, and ensure that these two events always occur in the correct sequence. Central to these controls is a class of enzymes called the cyclin dependent kinases (CDKs) which are important in regulating the cell cycle in all eukaryotic organisms, from yeast to humans. Much of what we know about these controls is derived from simple systems such as the yeasts and the frog *Xenopus*. For this reason after a brief introduction to CDKs, the controls in one well understood model system—the fission yeast—are described, before summarising present knowledge in mammalian cells.

Cyclin dependent kinases

This class of protein kinases consists of a 34 kilodalton (kDa) catalytic core which is associated with a cyclin partner. In simple eukaryotes, there is only a single gene encoding the 34 kDa catalytic core, the *cdc2* gene in fission yeast and the *CDC28* gene in budding yeast. In more complex metazoan eukaryotes, a whole family of CDKs has been identified including *CDK2*, -4, and -6, but not all family members are involved in cell cycle control. Several cyclin partners are found in the yeasts and, in the metazoa, many different types have been identified, such as cyclins A, B, and E. The different cyclins may influence the substrate specificity of the catalytic core, or other aspects of function such as location within the cell or the timing of protein kinase activation.

Three major modes of CDK regulation have been identified. The first is the association and dissociation of the two components, followed by destruction of the cyclin component. Association is required for the enzyme to have activity, and dissociation destroys activity. The second mode of regulation concerns phosphorylation changes within the active site of the enzyme, particularly the tyrosine residue Y15, but also the adjacent threonine T14. When phosphorylated, the enzyme activity is reduced and for full activity both residues must be dephosphorylated. The third mode of regulation involves another set of generally small protein molecules called CDK inhibitors, which, when associated with the CDK, substantially reduce its activity. These various modes of regulation influence CDK activity during the cell cycle, bringing about activation and inactivation of differing catalytic core/cyclin partners. The changes of CDK activity are thought to drive the cell into S phase and mitosis, and to ensure that these events occur in the correct sequence. How this operates in fission yeast will be discussed in the next section.

Fission yeast: onset of the S phase and mitosis

The best understood control in fission yeast is that regulating the onset of mitosis. Mitosis is promoted by a CDK consisting of the catalytic core p34^{cdc2} and a B cyclin partner p56^{cdc13} encoded by *cdc13*. The p56^{cdc13} B cyclin accumulates mostly during G2, and sufficient of the complex is made early during G2 for the cell to undergo mitosis. Restraint from full activation during the rest of G2 is brought about by the Y15 inhibitory phosphorylation. This results from an inhibitory *wee1* encoded protein kinase which is countered by an activatory *cdc25* encoded protein phosphatase. Thus the actual timing of protein kinase activation and mitotic onset is determined by the balance between the *wee1* and *cdc25* activities.

When the p34^{cdc2}/p56^{cdc13} protein kinase is fully activated, cells enter mitosis. A number of substrates have been identified which become phosphorylated by the protein kinase, and in metazoan cells these are thought to be involved in nuclear envelope breakdown, chromosome condensation, and the generation of the mitotic spindle. Once the events of mitosis have taken place, the p34^{cdc2}/p56^{cdc13} protein kinase activity is dramatically reduced by destroying the p56^{cdc13} B cyclin partner. As a consequence the cell exits mitosis and enters G1. The accumulation of p56^{cdc13} B cyclin during G2 and its destruction at the end of mitosis result in a periodic change in p56^{cdc13} level during the cell cycle. This type of behaviour is typical for the cyclins and accounts for their name, because they "cycle" in level through the cell cycle.

The control acting in G1 regulating the onset of the S phase also makes use of p34^{cdc2}, but this time the major B cyclin partner is encoded by *cig2*. This CDK complex appears during G1 and is required to promote onset of the S phase. The level of protein kinase activity necessary for the S phase is reduced compared with that required for mitosis. Total p34^{cdc2} protein kinase activity rises to a low level in late G1, remains at this low level during the S phase and much of G2, only rising to a high level in late G2 to bring about mitosis. The regulation of the p34^{cdc2}/p45^{cig2} protein kinase is likely to involve changes in p45^{cig2} synthesis and proteolysis, and the substrates are likely to include proteins required for DNA replication.

Fission yeast: ensuring the correct sequence

If either S phase or mitosis is blocked or delayed in fission yeast as in most eukaryotes, then the next event in the sequence is also delayed. This ensures genomic stability, because a cell that has not completed S phase would be unable to partition a full set of replicated chromosomes at mitosis whereas a cell that undergoes another S phase before mitosis takes place would undergo a change in ploidy. These phenomena are both examples of checkpoint controls, whereby a cell at a particular point in the cell cycle monitors or checks the occurrence of an earlier event, and if it is not properly completed a block is imposed on further cell cycle progression. A block within S phase prevents activation of the p34^{cdc2}/p56^{cdc13} protein kinase. One possible mechanism is that the cell monitors the presence of protein complexes required for the initiation of DNA replication, and if they are present then the inhibitory phosphorylation on Y15 is maintained. When S phase is completed these complexes disappear and so the signal maintaining the inhibitory Y15 phosphorylation is switched off, allowing the cell to proceed into mitosis. If a fission yeast cell is blocked earlier in G1, before the appearance of these complexes, then the p34^{cdc2}/p56^{cdc13} protein kinase is still inhibited but this time by a different mechanism. A CDK inhibitor p25^{rum1} appears in G1 cells and inhibits the p34^{cdc2}/p56^{cdc13} protein kinase, preventing entry into mitosis.

A cell in G2 does not normally re-initiate another S phase. This control also involves the p34^{cdc2}/p56^{cdc13} protein kinase. The low level of p34^{cdc2}/p56^{cdc13} protein kinase during G2 blocks initiation of another round of DNA replication. If all p34^{cdc2}/p56^{cdc13} protein kinase activity is destroyed, then the cell re-enters the S phase. Presumably, the G2 protein kinase activity phosphorylates and inhibits substrates that are necessary to initiate DNA replication. During the normal cell cycle, the destruction of the p34^{cdc2}/p56^{cdc13} protein kinase at the end of mitosis removes this block over the initiation of DNA replication allowing the S phase to take place. In this manner it is ensured that there is only one S phase per cell cycle.

S phase and mitosis controls in mammalian cells

The basic control scheme involving a central role for CDKs identified in simple eukaryotes and outlined above for fission yeast also applied to the metazoa, including mammalian cells. Onset of the S phase requires *CDK2* encoding p34^{CDK2}, partnered by cyclin E. The level of this complex and its associated protein kinase activity rises in late G1 and is present during the S phase. The p34^{CDK2}/cyclin E kinase is rate limiting for G1 progression because high level expression of cyclin E shortens G1, leading to an advancement of S phase. There may also be a role for p34^{CDK2} partnered by cyclin A for S phase, although this has not been so fully characterised. The role of p34^{CDK2} appears to be equivalent to that played by p34^{cdc2} in the G1 to S phase transition in fission yeast, and indeed mammalian *CDK2* can substitute for *cdc2* at this point of the cell cycle.

Once into G2 a complex is formed between $p34^{CDC2}$ and cyclin B. This is phosphorylated on the Y15 residue of $p34^{CDC2}$ and so has low protein kinase activity. There are metazoan homologues of *wee1* and *cdc25* which are important for controlling the level of Y15 phosphorylation. As in fission yeast this residue becomes dephosphorylated in late G2, and the protein kinase activity rises to a high level to bring about mitosis. This increase is rate limiting for mitotic onset because introducing active $p34^{CDC2}$/cyclin B into cells advances them into mitosis. The cyclin B component is degraded at the exit of mitosis as the $p34^{CDC2}$/cyclin B protein kinase activity drops to a low level and cells re-enter G1.

This pattern of CDK protein kinase changes is clearly reminiscent of that described in fission yeast and suggests that an increasing level of activity is associated with progression through the cell cycle and the onset of both the S phase and mitosis. However, the increased diversity of the complexes involved, including $p34^{CDC2}$, $p34^{CDK2}$, and cyclins A, B, and E, suggests that there may be differing substrate specificities associated with the various cell cycle transitions. Certainly this diversity allows differing regulation at the various stages of the cell cycle. There is some evidence that the CDKs in metazoa are also important for ensuring the correct sequence of the S phase and mitosis during the cell cycle. When S phase is blocked, $p34^{CDC2}$/cyclin B activity is restrained and mitosis inhibited, whereas introduction of active protein kinase into blocked cells overcomes the restraint blocking mitosis. This may work through Y15 phosphorylation. It is not clear if $p34^{CDC2}$/cyclin B represses re-initiation of DNA replication during G2, but in endoreduplicating cells undergoing repeated S phases, this protein kinase is at a low level. Interestingly, $p34^{CDC2}$/cyclin B protein kinase activity may not drop to such a low level in between the two nuclear divisions during meiosis, perhaps explaining why there is no second S phase during the meiotic cell cycle.

CDKs and growth control

When the cells are not actively undergoing reproduction they exit from the cell cycle and enter a quiescent state called G0. The shift between the quiescent state and growth plays an important part in the development and maintenance of the organism, being required for such processes as wound healing and tissue replenishment. If regulation of this shift is defective it may result in uncontrolled cell reproduction, leading to cancer. It has become clear that CDKs are also involved in the control by regulating the shift from quiescence to growth in mammalian cells. Both *CDK4* and *-6* are partnered by cyclin D function during G1, where they phosphorylate the tumour suppressing RB (retinoblastoma) protein allowing release of the E2F transcription factor responsible for transcribing some of the genes required for the G1 to S phase transition. These CDKs are also regulated by CDK protein inhibitors including the p15/p16 INK4 family and the p21/p27 WAF1/KIP family. Interestingly, a number of these components have been implicated in the changes that occur during cancer formation and tumorigenesis. Examples include the Rb protein in retinoblastoma, cyclin D in breast tumours, and p16 in melanomas. This regulatory pathway plays an important role in growth control, acting at the point of entry into the cell cycle.

Thus, the CDKs are central to cell cycle controls in all eukaryotic organisms and, in the metazoa, their function has further diversified to include involvement in growth control. Understanding their mode of action will therefore be essential both for a full knowledge of cell reproduction and development, and for an understanding of cancer.

Further reading

Murray A, Hunt T. *The cell cycle: an introduction.* New York: WH Freeman & Co., 1993.

Nurse P. Universal control mechanism regulating onset of M-phase. *Nature* 1990; 344:503–8.

Nurse P. Ordering S phase and M phase in the cell cycle. *Cell* 1994;79:547–50.

Sherr C, Roberts J. Inhibitors of mammalian G1 cyclin-dependent kinases. *Genes Devel* 1995;9:1149–63.

7: Apoptosis

Mary K L Collins

Apoptosis was first described by pathologists who noted morphological characteristics of dying cells in tissue sections.[1] These included cytoplasmic shrinkage, plasma membrane blebbing, DNA condensation, and nuclear fragmentation. Apoptosis, derived from a Greek word for shedding of leaves, is now thought to be the physiological mechanism by which all cells die. Therefore, the regulation of apoptosis is as important as the regulation of cell proliferation or differentiation in controlling tissue size, structure, and function.

The common molecular mechanism

Apoptosis is probably general to all muticellular organisms. Indeed, the study of the control of apoptosis in worms, where genetic approaches are possible, has identified regulatory molecules which also control apoptosis in mammalian cells. The Bcl-2 protein is one such conserved inhibitor of apoptosis.[2] Its mechanism of action remains unknown; it is found associated with intracellular membranes and is a member of a family of related proteins, some of which stimulate and some of which inhibit apoptosis when over-expressed.[3] Downstream of Bcl-2 in the common apoptotic pathway (fig 7.1) is a family of proteases, the ICE like proteases,[4] so called because the first member to be identified was interleukin-1β converting enzyme. Activation of some or all of these proteases triggers the final irreversible death of the cell. At this stage the collapse of the cell nucleus is precipitated by ICE like protease digestion of proteins such as the lamins which strengthen the nuclear envelope. An endonuclease then digests the DNA.[5] The dying cell changes its surface and is then recognised and ingested by one of its neighbours.[6] This ensures that the debris is rapidly cleared, which prevents an inflammatory or immune response.

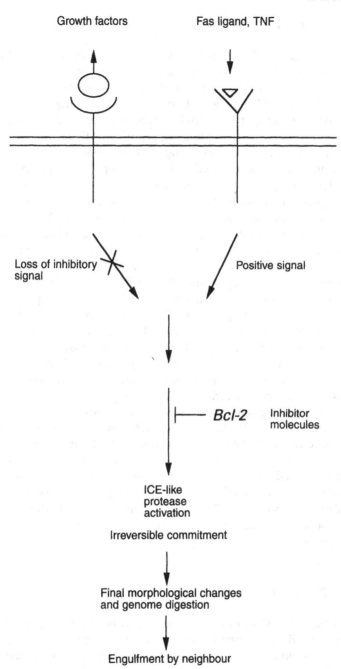

FIG 7.1—*The common mechanism of apoptosis.*

Role of apoptosis

As apoptotic cells are rapidly cleared in tissues, the extent of apoptosis during development and tissue turnover has often been greatly underestimated. Martin Raff recently proposed that a large proportion of cells that are "born" in all developing tissues are destined to enter apoptosis rapidly.[7] This allows tissue development to be controlled by a limiting supply of growth factors, which inhibit the apoptotic pathway (fig 7.1) and rescue appropriately positioned and differentiated cells from apoptosis. Regulation of cell death also provides a powerful mechanism to select appropriate subpopulations of cells, for example, in the immune system. In the thymus, developing self reactive T lymphocytes, which would be potentially harmful, are induced to enter apoptosis.[8] Some mature T lymphocytes responding to antigen are also eliminated by apoptosis, providing a mechanism for switching off an immune response.[9] Antigen triggering of these cells results in the synthesis of the Fas ligand which occupies the Fas surface receptor and stimulates autocrine apoptosis.[10] Tumour necrosis factor (TNF) also stimulates apoptosis in mature T cells;[11] its receptor shares some regions of homology with Fas, suggesting that the two may initiate a common signalling pathway (fig 7.1). In lymph nodes, B lymphocytes that recognise foreign antigens are rescued from apoptosis, allowing selection of high affinity antibodies,[12] whereas those recognising self antigens are deleted by apoptosis[13] in a Fas dependent manner.[14] Fas ligand is highly expressed in sites such as the testis[15] and eye[16] so that immune cells entering these sites are killed, because an inflammatory response leading to tissue damage would be undesirable. Apoptosis is also used by cytotoxic T cells to kill virally infected cells. They trigger apoptosis in their targets both through the Fas pathway[17] and by the introduction of a cytotoxic protease into the target which directly activates ICE like proteases.[18]

Apoptosis in disease

Bcl-2 was first identified as a gene product overexpressed in a subset of B cell lymphomas[19] which implies that inhibition of apoptosis can contribute to tumorigenesis. Bcl-2 overexpression in B cells of transgenic mice results in larger numbers of B cells,[20] but the oncogene *c-myc* needs to be co-expressed to drive cell proliferation and allow malignant tumours to form.[21] This potent cooperation between myc and Bcl-2 can be explained by the fact that overexpression of c-myc alone causes cells to enter apoptosis,[22] which is prevented by coexpression of Bcl-2.[23] Myc and Bcl-2 functions are probably upregulated in a number of human tumours, for example, in prostate cancer Bcl-2 is expressed at high levels in androgen independent tumours[24] and the gene encoding Mxi1, a negative regulator of Myc, is mutated in some prostate tumours.[25]

The p53 protein functions to suppress tumour formation; p53 is frequently mutated in human tumours[26] and inheritance of a mutated p53 allele predisposes individuals to a variety of cancers.[27 28] Mutant p53 probably acts to block the function of wild type p53,[29 30] as mice homozygously deleted for p53 show the mutant phenotype of enhanced tumour formation.[31] p53 acts, at least in part, by regulating gene expression. Upon binding a specific DNA sequence it transactivates target genes;[32] it also represses transcription of genes lacking a p53 target sequence by interacting with the basal transcription machinery.[33] Cells are often induced to enter apoptosis after ionising radiation[34] or other forms of DNA damage. In some cell populations such as thymocytes,[35 36] but not mature lymphocytes,[37] this radiation induced apoptosis is p53 dependent. p53 has also been reported to be required for apoptosis induced by dysregulation of expression of the cell cycle regulators such as c-Myc.[38] As both p53 mutation and Bcl-2 overexpression[39] block apoptosis induced by DNA damage, many tumours are intrinsically somewhat resistant to conventional tumour therapies that involve DNA damage by radiation or cytotoxic drugs. Resistance to apoptosis probably also allows tumour cells to survive in hostile, for example, hypoxic, conditions.[40]

Mice that are defective in Fas[41] or Fas ligand genes[42] have an autoimmune disease not unlike systemic lupus erythematosus. This suggests that patients with systemic autoimmune disease may have defects in apoptosis; in a small number of cases this can be shown to result from a genetic defect in Fas.[43 44] Defects in Fas cause autoimmunity because Fas triggering leading to apoptosis is the mechanism by which T and B cells that recognise self antigens are eliminated (see above). Apoptotic cell debris is immunogenic, as demonstrated by the detection of autoantibodies directed against apoptotic cell structures in patients with systemic lupus erythematosus.[45] It is therefore also possible that aberrant apoptosis induced by infection (see below) or tissue damage may be involved in the generation of tissue specific or systemic autoimmune disease. Tissue damage by induction of apoptosis is likely to follow many medical procedures which result in brief periods of ischaemia.[46]

Apoptosis in viral infection

Viral infection can result in both "direct" cell damage, whereby the virally infected cell dies as a result of cytopathicity of the virus, and "indirect" pathogenesis after an immune attack on infected cells. It has long been recognised that features of the cytopathic effect (CPE) induced by many viruses resemble those of apoptosis. Recent studies have demonstrated that several viruses can specifically trigger the apoptotic pathway, including influenza,[47] adenovirus,[48] and HIV.[49] In some studies the mechanism of viral induction of apoptosis has been investigated by demonstrating that expression of a single viral gene product is effective, for example, GP120, the envelope protein of the HIV type 1 (HIV 1) triggers apoptosis when it binds to the viral receptor CD4; GP120 and the viral TAT protein upregulate Fas ligand.[50] It has been suggested that this might explain the severe T cell depletion induced by HIV; however, T cell apoptosis is a feature of the immune response to other viral infections.[51] Viral stimulation of apoptosis will result in more rapid death of the infected cell, which will limit the extent of virus replication. Indeed, overexpression of Bcl-2 in target cells has been shown to inhibit lysis after Sindbis viral infection and to allow latency.[52] This is an important observation, because it demonstrates that target cell status can determine the outcome of viral infection.

Viruses themselves can also inhibit apoptosis in infected cells, thereby controlling the rate of target cell death, to increase virus replication. Figure 7.2 shows that viruses encode a variety of gene products which inhibit the apoptotic pathway at a number of steps. Three highly diverse viruses encode bcl-2 homologues.[48 53 54] Mutant adenoviruses lacking the E1B homologue show decreased ability to replicate, which can be restored by overexpressing bcl-2 in the target cell.[55] The cowpox crmA gene product and the baculovirus p35 protein are ICE like protease inhibitors.[56 57] CrmA overexpression inhibits apoptosis in neurons deprived of growth factor which therefore implicates such a protease in the regulation of neuronal cell death.[58] Baculoviruses from which the p35 gene has been deleted produce less progeny in culture and are less infectious

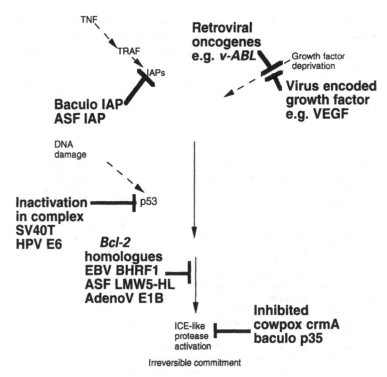

FIG 7.2—*Viral proteins that inhibit apoptosis.*

to insects.[59] Baculoviruses and African swine fever virus also encode a further anti-apoptotic regulator IAP.[60] Cellular homologues of IAP are associated with the TNF receptor[61] and mutation of the gene for one of these is thought to result in neurodegenerative disease.[62] The role of other viral gene products that substitute for components on the growth signalling pathway,[63] or complex with p53,[64] may be in the promotion of infected cell growth as well as the suppression of infected cell apoptosis. An important component of tissue damage after viral infection will clearly be what follows immune stimulation by the foreign viral antigens, which can result in an accumulation of a variety of inflammatory cells such as macrophages, granulocytes, and eosinophils at a site of immune stimulation. The immune response to viral infection may provide some explanation for the induction of apoptosis by viruses. As apoptosis results in the clearance of cellular debris, the virally infected cell dying by apoptosis will not stimulate the immune system to the same extent as a lysed cell releasing its cytoplasmic and nuclear contents. An immune response against intracellular antigens will lead to an increased antiviral response by analogy with the adjuvant principle. Thus, regulation of apoptosis by viruses will be a balance between the optimisation of virus production by the inhibition of apoptosis and the minimisation of immune response by the induction of apoptosis.

1 Kerr J, Wyllie A, Currie G. Apoptosis: a basic biological phenomenon with wide ranging implications in tissue kinetics. *Br J Cancer* 1972;26:239–57.
2 Hengartner MO, Horvitz HR. *C. elegans* cell survival gene ced-9 encodes a functional homolog of the mammalian proto-oncogene bcl-2. *Cell* 1994;76: 665–76.
3 Korsmeyer SJ. Regulators of cell death. *Trends Genet* 1995;11:101–5.
4 Yuan J, Shaham S, Ledoux S, Ellis HM, Horvitz HR. The *C. elegans* cell death gene ced-3 encodes a protein similar to mammalian interleukin-1β-converting enzyme. *Cell* 1993;75:641–52.
5 Wyllie AH. Glucocorticoid-induced thymocyte apoptosis is associated with endogenous endonuclease activation. *Nature* 1980;284:555–6.
6 Savill J, Fadok V, Henson P, Haslett C. Phagocyte recognition of cells undergoing apoptosis. *Immunol Today* 1993;14:131–6.
7 Raff MC. Social controls on cell survival and cell death. *Nature* 1992;356: 397–400.
8 Murphy KM, Heimberger AB, Loh DY. Induction by antigen of intrathymic apoptosis of CD4 + CD8 + TCRlo thymocytes in vivo. *Science* 1990;250:1720–3.
9 Critchfield J, Racke M, Zuniga-Pflucker J, Canella B, Raine C, Goverman J, *et al.* T cell deletion in high antigen dose therapy of autoimmune encephalomyelitis. *Science* 1994;263:1139–43.
10 Dhein J, Walczak H, Baumler C, Debatin K, Krammer P. Autocrine T-cell suicide mediated by APO-1. *Nature* 1995;373:438–41.
11 Zheng L, Fisher G, Miller R, Peschon J, Lynch D, Lenardo M. Induction of apoptosis in mature T cells by tumour necrosis factor. *Nature* 1995;377:348–51.
12 Liu YJ, Joshua DE, Williams GT, Smith CA, Gordon J, MacLennan IC. Mechanism of antigen-driven selection in germinal centres. *Nature* 1989;342: 929–31.
13 Pulendran B, Kannourakis G, Nouri S, Smith K, Nossal G. Soluble antigen can cause enhanced apoptosis of germinal centre B cells. *Nature* 1995;375: 331–4.
14 Rathmell J, Cooke M, Ho W, Grein J, Townsend S, Davis M, Goodnow C. Fas-dependent elimination of self-reactive B cells upon interaction with CD4 + T cells. *Nature* 1995;376:181–4.
15 Bellgrau D, Gold D, Selawry H, Moore J, Franzusoff A, Duke R. A role for CD95 ligand in preventing graft rejection. *Nature* 1995;19:630–2.
16 Griffith T, Brunner T, Fletcher S, Green D, Ferguson T. Fas ligand-induced apoptosis as a mechanism of immune privilege. *Science* 1995;270:1189–92.
17 Golstein P. Fas-based T cell-mediated cytotoxicity. *Curr Top Microbiol Immunol* 1995;198:25–37.
18 Darmon A, Nicholson D, Bleackley R. Activation of the apoptotic protease CPP32 by cytotoxic T-cell-derived granzyme B. *Nature* 1995;377:446–8.
19 Tsujimoto Y, Croce C. Analysis of the structure, transcripts and protein products of bcl-2, the gene involved in human follicular lymphoma. *Proc Natl Acad Sci USA* 1986;83:5214–18.

20 McDonnell T, Deane N, Platt F, Nunez G, Jaeger U, McKearn J, et al. Bcl-2-immunoglobulin transgenic mice demonstrate extended B cell survival and follicular lymphoproliferation. Cell 1989;57:79–88.

21 Strasser A, Harris AW, Bath ML, Cory S. Novel primitive lymphoid tumours induced in transgenic mice by cooperation between myc and bcl-2. Nature 1990;348:331–3.

22 Evan GI, Wyllie AH, Gilbert CS, Littlewood TD, Land H, Brooks M, et al. Induction of apoptosis in fibroblasts by c-myc protein. Cell 1992;69:119–28.

23 Fanidi A, Harrington EA, Evan GI. Co-operative interaction between c-myc and bcl-2 proto-oncogenes. Nature 1992;359:554–6.

24 McDonnell T, Troncoso P, Brisbay S, Logothetis C, Chung L, Hsieh J, et al. Expression of the protooncogene bcl-2 in the prostate and its association with emergence of androgen-independent prostate cancer. Cancer Res 1992;52: 6940–4.

25 Eagle L, Yin X, Brothman A, Williams B, Atkin N, Prochownik E. Mutation of the MX11 gene in prostate cancer. Nat Genet 1995;9:249–55.

26 Hollenstein M, Sidransky D, Vogelstein B, Harris C. p53 mutations in human cancers. Science 1991;253:49–53.

27 Srivastava S, Zou Z, Pirollo K, Blattner W, Chang E. Germ-line transmission of a mutated p53 gene in a cancer-prone family with Li–Fraumeni syndrome. Nature 1990;348:747–9.

28 Malkin D, Li F, Strong L, Fraumeni J, Nelson C, Kim D, et al. Germ line p53 mutations in a familial syndrome of breast cancer, sarcomas and other neoplasms. Science 1990;250:1233–8.

29 Herskowitz I. Functional inactivation of genes by dominant negative mutations. Nature 1987;329:219–22.

30 Shaulian E, Zauberman A, Ginsberg D, Oren M. Identification of a minimal transforming domain of p53. Mol Cell Biol 1992;12:5581–92.

31 Donehower L, Harvey M, Slagle B, McArthur M, Montgomery C, Butel J, et al. Mice deficient for p53 are developmentally normal but susceptible to spontaneous tumours. Nature 1992;356:215–21.

32 Kern S, Kinzler K, Bruskin A, Jarosz D, Friedman P, Prives C, et al. Identification of p53 as a sequence specific DNA-binding protein. Science 1991;252:1708–11.

33 Seto E, Usheva A, Zambetti G, Momand J, Horikoshi N, Weinmann R, et al. Wild-type p53 binds to the TATA-binding protein and represses transcription. Proc Natl Acad Sci USA 1992;89:12028–32.

34 Neal J, Potten C. Effect of low dose ionizing radiation on the murine pericryptal fibroblast sheath. Int J Radiat Biol Relat Stud Phys Chem Med 1981;39:175–85.

35 Clarke A, Purdie C, Harrison D, Morris R, Bird C, Hooper M, et al. Thymocyte apoptosis induced by p53-dependent and independent pathways. Nature 1993; 362:849–52.

36 Lowe SW, Schmitt EM, Smith SW, Osborne BA, Jacks T. p53 is required for radiation induced apoptosis in mouse thymocytes. Nature 1993;362:847–9.

37 Strasser A, Harris A, Jacks T, Cory S. DNA damage can induce apoptosis in proliferating lymphoid cells via p53-independent mechanisms inhibited by Bcl-2. Cell 1994;79:329–39.

38 Hermeking H, Eick D. Mediation of C-Myc-induced apoptosis by p53. Science 1994;265:2091–3.

39 Collins MKL, Marvel J, Malde P, Lopez-Rivas A. Interleukin 3 protects murine bone marrow cells from apoptosis induced by DNA damaging agents. J Exp Med 1992;176:1043–51.

40 Graeber T, Osmanian C, Jacks T, Housman D, Koch C, Lowe S, et al. Hypoxia-mediated selection of cells with diminished apoptotic potential in solid tumours. Nature 1996;379:88–91.

41 Watanabe-Fukunaga R, Brannan CI, Copeland NG, Jenkins NA, Nagata S. Lymphoproliferation disorder in mice explained by defects in fas antigen that mediates apoptosis. Nature 1992;356:314–17.

42 Takahashi T, Tanaka M, Brannan C, Jenkins N, Copeland N, Suda T, et al. Generalised lymphoproliferative disease in mice caused by a point mutation in the fas ligand. Cell 1994;76:969-76.

43 Fisher G, Rosenberg F, Straus S, Dale J, Middleton L, Lin A, et al. Dominant interfering Fas gene mutations impair apoptosis in a human autoimmune lymphoproliferative disease. Cell 1995;81:935-46.

44 Rieux-Laucat F, Le-Deist F, Hivroz C, Roberts I, Debatin K, Fischer A, et al. Mutations in Fas associated with human lymphoproliferative syndrome and autoimmunity. Science 1995;268:1347-9.

45 Casciola-Rosen L, Anhalt G, Rosen A. Autoantigens targeted in systemic lupus erythematosus are clustered in two populations of surface structures on apoptotic keratinocytes. J Exp Med 1994;179:1317-30.

46 Schumer M, Colombel M, Sawczuk I, Gobe G, Connor J, O'Toole K, et al. Morphological, biochemical and molecular evidence of apoptosis during reperfusion phase after brief periods of renal ischemia. Am J Pathol 1992;140: 831-8.

47 Hinshaw VS, Olsen CW, Dybdahl-Sissoko N, Evans D. Apoptosis: a mechanism of cell killing by influenza A and B viruses. J Virol 1994;68:3667-73.

48 Rao L, Debbas M, Sabbatini P, Hockenberry D, Korsmeyer S, White E. The adenovirus E1A proteins induce apoptosis, which is inhibited by the E1B19-kDa and Bcl-2 proteins. Proc Natl Acad Sci USA 1992;89:7742-6.

49 Schwarz O, Alizon M, Heard JM, Danos O. Impairment of T cell receptor-dependent stimulation in CD4+ lymphocytes after contact with membrane-bound HIV-1 envelope glycoprotein. Virology 1994;198:360-5.

50 Westendorp M, Frank R, Ochsenbauer C, Stricker K, Dhein J, Walczak H, et al. Sensitisation of T cells to CD95 mediated apoptosis by HIV-1 TAT and gp120. Nature 1995;375:497-500.

51 Akbar AN, Borthwick N, Salmon M, Gombert W, Bofill M, Shamsadeen N, et al. The significance of low bcl-2 expression by CD45RO T cells in normal individuals and patients with acute viral infections. The role of apoptosis in T cell memory. J Exp Med 1993;178:427-38.

52 Levine B, Huang Q, Isaacs JT, Reed JC, Griffin DE, Hardwick JM. Conversion of lytic to persistent alphavirus infection by the Bcl-2 oncogene. Nature 1993; 361:739-42.

53 Neilan JG, Lu Z, Afonso CL, Kutish GF, Sussman MD, Rock DL. An African swine fever virus gene with similarity to the proto-oncogene bcl-2 and the Epstein-Barr virus gene BHRF1. J Virol 1993;67:4391-4.

54 Henderson S, Huen D, Rowe M, Dawson C, Johnson G, Rickinson A. Epstein-Barr virus-coded BHRF1 protein, a viral homologue of Bcl-2, protects human B cells from programmed cell death. Proc Natl Acad Sci USA 1993;90: 8479-83.

55 Chiou SK, Tseng CC, Rao L, White E. Functional complementation of the adenovirus E1B protein with Bcl-2 in the inhibition of apoptosis in infected cells. J Virol 1994;68:6553-66.

56 Xue D, Horvitz H. Inhibition of the C. elegans cell-death protease CED-3 by a CED-3 cleavage site in baculovirus p35 protein. Nature 1995;377:248-51.

57 Bump N, Hackett M, Huginin M, Seshagiri S, Brady K, Chen P, et al. Inhibition of ICE family proteases by baculovirus antiapoptotic protein p35. Science 1995; 269:1885-8.

58 Gagliardini V, Fernandez P-A, Lee RKK, Drexler HCA, Rotello RJ, Fishman MC, et al. Prevention of neuronal death by the crmA gene. Science 1994;263: 826-8.

59 Clem RJ, Miller LK. Apoptosis reduces both the in vitro replication and the in vivo infectivity of a baculovirus. J Virol 1993;67:3730-8.

60 Birnbaum MJ, Clem RJ, Miller LK. An apoptosis-inhibiting gene from a nuclear polyhidrosis virus encoding a polypeptide with Cys/His motifs. J Virol 1994;68: 2521-8.

61 Rothe M, Pan M, Henzel W, Ayres T, Goeddel D. The TNFR2-TRAF signalling complex contains two novel proteins related to baculoviral inhibitor of apoptosis proteins. *Cell* 1995;**83**:1243-52.
62 Liston P, Roy N, Tamai K, Lefebvre C, Baird S, Cherton-Horvat G, *et al.* Suppression of apoptosis in mammalian cells by NAIP and a related family of IAP genes. *Nature* 1996;**379**:349-53.
63 Evans CA, Owen-Lynch PJ, Whetton AD, Dive C. Activation of Abelson tyrosine kinase activity is associated with suppression of apoptosis in haematopoietic cells. *Cancer Res* 1993;**53**:1735-8.
64 McCarthy SA, Symonds HS, Dyke TV. Regulation of apoptosis in transgenic mice by simian virus 40 T antigen-mediated inactivation of p53. *Proc Natl Acad Sci USA* 1994;**91**:3979-83.

8: Cell to cell and cell to matrix adhesion

D R Garrod

Maintenance of the structure and organisation of body tissues is dependent on the adhesion of cells to each other and to the extracellular matrix. In simple epithelia, such as the intestinal mucosa and the kidney tubule, individual cells have surfaces with three distinct sets of adhesive properties. The apical or luminal surface is non-adhesive, whereas the lateral and basal surfaces are specialised, respectively, for adhesion to each other and to the basement membrane (fig 8.1). In stratified epithelia, such as epidermis, the basal cells adhere both to the basement membrane and to each other. To move into the suprabasal layers, cells lose matrix adhesion and adhesion is lost altogether when outer layer cells are sloughed off.

Other cell types—leucocytes and blood platelets—spend much of their time circulating freely and thus showing no adhesive interactions. Lymphocytes, however, show quite specific recirculation patterns in which particular subsets leave the blood circulation by first adhering to high endothelial venule cells at specific sites—for example, peripheral lymph nodes or mucosal associated lymphoid tissue. They then migrate into the lymphoid tissue and eventually return to the blood circulation. Endothelial adhesion by leucocytes is also the first step in a range of adhesive interactions required in their tissue invasive response to inflammation (fig 8.2). Blood platelets respond to injury by deploying a whole set of adhesive interactions with endothelial cells, with the matrix of the clot and endothelial basement membrane, and with each other.[1]

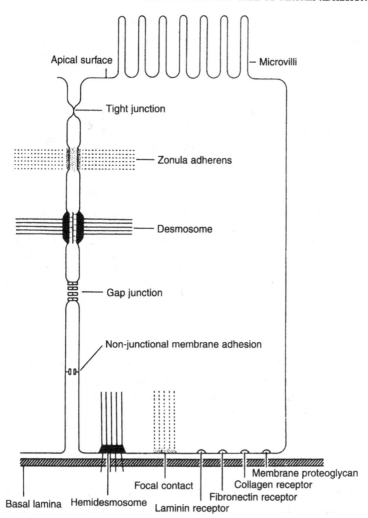

FIG 8.1—*Generalised simple epithelial cell showing arrangement of adhesion mechanisms. The apical surface is non-adhesive. The lateral surface mediates cell–cell interactions via adhesive junctions, the zonula adherens and the desmosome, and communicating gap junctions that allow small molecules to be exchanged between the cytoplasm of adjacent cells. The tight junction or zonula occludens restricts and regulates paracellular permeability. The basal surface mediates adhesion to the basement membrane. Hemidesmosomes are present in some simple epithelia but are most numerous in the epidermis (see fig 8.4). Integrin and non-integrin receptors for several matrix components are present here. When cells are cultured, many of these become localised in focal contacts, small areas where adhesion to the substratum is the strongest. (Reproduced from* Journal of Cell Science *by permission of the Company of Biologists.)*

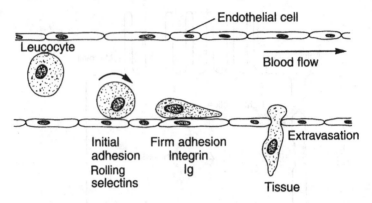

FIG 8.2—*Leucocyte adhesion—transendothelial migration. Three steps are involved in leucocyte extravasation during inflammation or infection. First, selectin mediated adhesion permits initial interactions in which the leucocyte rolls along the capillary wall, driven by the blood flow. Second, signals generated during rolling increase the affinity of integrin binding to immunoglobulin like molecules generating firm adhesion—the leucocyte stops rolling. Third, the leucocyte moves between capillary endothelial cells into the tissue.*

Thus cells of any particular type possess a set of adhesive properties which may be both spatially and temporally regulated. A considerable amount is now known about the molecular mechanisms that mediate these properties.[12]

Families of cell adhesion molecules

Many different cell adhesion molecules have been described. Most of them belong to one of a small number of families of related molecules in which the individual members share the same basic molecular structure, but are subtly different from each other (fig 8.3).

Cadherins

In most tissues a major contribution to intercellular adhesion is made by calcium dependent cell adhesion molecules known as cadherins.[3] In general these are simple transmembrane glycoproteins. The extracellular domain has an adhesion site towards the *N*-terminal region and several calcium binding sites.

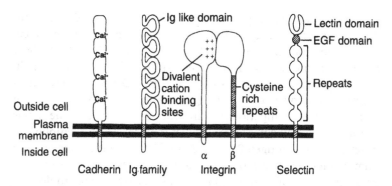

FIG 8.3—*Molecular structures of cell adhesion molecule families.*

Adhesive binding is mainly homophilic: a cadherin molecule on one cell binds to another cadherin molecule of the same type on the next cell. Linkage of the cytoplasmic domain to the cytoskeleton through proteins known as catenins is necessary for cadherin function. The best characterised is epithelial cadherin, E-cadherin. This appears very early in development when it is involved in compaction of the eight cell embryo and cell polarisation. In adult epithelia—for example, intestinal epithelium—it is present on the lateral cell surfaces but is concentrated in intercellular junctions, known as the zonulae adherentes, which ring the apicolateral margins of cells. These junctions are characterised by a cortical ring of cytoskeleton, the major component of which is actin. Cadherins are also present in the nervous system where they play an essential role in neural development.

The adhesive glycoproteins of the other major intercellular junctions of epithelia, the desmososmes, are also members of the cadherin family.[4] Their extracellular domains are very like those of cadherin but their cytoplasmic domains differ, being specialised for forming desmosomal plaques and, thereby, attachment to the keratin intermediate filament cytoskeleton rather than to actin.

Immunoglobulin superfamily

Another major group of cell to cell adhesion molecules are members of the immunoglobulin superfamily.[5] Their extracellular portions are characterised by the presence of at least one, and usually multiple, immunoglobulin like domains. Included in this group are several nervous system adhesion molecules such as the neural cell adhesion molecule (N-CAM), L1, and TAG, which are involved in neuronal guidance and fasciculation. Several members of the immunoglobulin family are concerned with antigen recognition and adhesion in T lymphocytes. These include the T cell receptor (CD3) and its co-receptors CD4 and CD8, which together recognise the following: complexes of antigen peptide and major histocompatibility complexes on other cells; the major histocompatibility complex molecules themselves; and lymphocyte function related antigen 2 (LFA-2 or CD2), a receptor for another immunoglobulin like molecule, LFA-3, expressed on other cells. Another group of immunoglobulin like cell adhesion molecules includes the so called intercellular adhesion molecules, ICAM-1, -2, and -3 which are more widely expressed, for example, on epithelial and endothelial cells, and V-CAM on endothelial cells. These are involved in the inflammatory response.

The immunoglobulin superfamily is large and diverse, probably because the basic structure of the immunoglobulin domain is versatile and readily adaptable to different binding functions. Among these molecules, however, only the T cell receptor and the immunoglobulins themselves have somatically variable domains necessary for antigen recognition. Members of the superfamily are present in insects, where they are involved in nerve connections; the association of immunoglobulin like domains in cellular recognition preceded the immune system in evolution.

Integrins

Both cell to cell and cell to matrix receptors are contained within this large family of adhesion molecules. Integrins are heterodimers consisting of one α chain and one β chain, both of which are necessary for adhesive binding. Sixteen different α chains and eight different β chains are now known. Integrins may be classified into subfamilies according to which β subunit is involved in the complex. Thus β1 integrin may associate with one of nine different α subunits to give a series of matrix receptors of differing specificity. The β2 integrins, on the other hand, are a family of cell to cell adhesion molecules of lymphoid cells with three alternative α subunits. The classification is made more complicated because some α subunits can associate with different β subunits (for example, α6β1 and α6β4).

Some integrins are apparently quite specific in their ligand binding properties—for example, α5β1 for the arginine, glycine, and aspartic acid (-Arg-Gly-Asp) tripeptide sequence of fibronectin—whereas others are promiscuous—for example, $\alpha v \beta 3$, once regarded as the vitronectin receptor, also binds fibronectin, fibrinogen, von Willebrand's factor, thrombospondin, and osteopontin. An interesting example is $\alpha 4 \beta 1$, which binds both the IIICS domain of fibronectin and V-CAM on endothelial cells. To complicate matters further, individual cell types usually express multiple integrins. A good example to consider here is the blood platelets that express predominantly $\alpha IIb \beta 3$ (GPIIb/IIIa), which binds fibrinogen, fibronectin, von Willebrand's factor, and vitronectin, but also lesser amounts of $\alpha V \beta 3$, $\alpha 5 \beta 1$, $\alpha 2 \beta 1$ (collagen), and $\alpha 6 \beta 1$ (laminin).

Selectins

Most cellular interactions seem to entail homophilic or heterophilic protein to protein binding. However, the selectins contribute a family of cell adhesion molecules that bind to carbohydrate. Selectins have lectin like domains at their extracellular N-terminal extremities. One of these, L-selectin (L= leucocyte) is a "homing receptor", mediating regionally specific adhesion of lymphocytes to endothelium in peripheral lymph nodes. This molecule is also involved in the adhesion of neutrophils to endothelium during the inflammatory response. Two other members of this family, E-selectin (E = endothelial) and P-selectin (P = platelet), also participate in the inflammatory response. E-selectin is upregulated on endothelial cells over a period of hours after stimulation by inflammatory mediators. P-selectin is contained within Wiebel–Palade bodies of endothelial cells and platelet α granules from which it is rapidly mobilised on activation, mediating adhesion to neutrophils and monocytes.

Cell adhesion and disease

Abnormalities of cell adhesion are involved in several diseases (table 8.1).[11-14] Adhesion defects may be broadly divided into two types: those in which adhesion is decreased because of inherited or acquired genetic defects in adhesion mechanisms, or because of disruption of adhesions by autoantibodies; and those in which adhesion is increased as in inflammatory diseases and atherosclerosis.

TABLE 8.1—*Diseases involving cell adhesion*

	Cell type or tissue affected	Molecule(s) affected or involved
Inherited diseases		
Leucocyte adhesion deficiency I	Leucocytes	β2 Integrins
Leucocyte adhesion deficiency II	Leucocytes	Sialyl Lewis X (ligand for E-and P-selectin)
Glanzmann's thrombasthenia	Blood platelets	GPIIb/IIIa integrin
Bernard–Soulier syndrome	Blood platelets	GPIb/IX (membrane receptor for von Willebrand's factor)
von Willebrand's disease	Blood platelets	von Willebrand's factor
Epidermolysis bullosa		
Junctional	Dermoepidermal junction	Laminin 5/α6β4 integrin
Dystrophic	Dermoepidermal junction	Collagen VII
Simplex	Basal epidermal keratinocytes	Keratin 5/14
Autoimmune diseases		
Pemphigus vulgaris (PV)	Epidermal keratinocytes	Desmoglein 3 (PV antigen)
Pemphigus foliaceus (PF)	Epidermal keratinocytes	Desmoglein 1 (PF antigen)
Bullous pemphigoid (BP)	Dermoepidermal junction	Collagen XVII (BP antigen 2)
Inflammatory diseases		
Ischaemic damage, rheumatoid arthritis, graft rejection, asthma, psoriasis, atopic dermatitis.	Leucocytes–endothelial cells	Selectins; α4β1 integrin—V-CAM; αLβ2 integrin—ICAM
Atherosclerosis	Blood platelets	Platelet adhesion receptors especially GPIIb/IIIa
Cancer		
Invasion and metastasis	Various	Integrins, E-cadherin, desmosomes, CD44

Adhesion of leucocytes (neutrophils) to capillary endothelium is an essential step in the acute inflamatory response. Genetic defects in β2 integrin or the fucosylated sialyl Lewis X blood group antigen, the carbohydrate ligand for E- and P-selectins, give rise to the leucocyte adhesion deficiencies LAD-I and LAD-II. Patients with these deficiencies have recurrent infection—and high mortality in LAD-I. The defect in LAD-II affects initial, rolling adhesion of leucocytes to endothelium whereas the β2 integrin defect in LAD-I affects the second, firm phase of leucocyte adhesion (see fig 8.2).

Chronically enhanced leucocyte–endothelial cell adhesion is a major contributor to inflammatory disease, in which invading leucocytes cause tissue damage. A common feature in enhancement of leucocyte adhesion is the continuous expression of endothelial cell adhesion molecules which are normally only transiently expressed in acute inflammation.

Reduced adhesion of blood platelets results from genetic defects in the adhesion receptors integrin IIb/IIIa and non-integrin Ib/IX, as well as in the extracellular ligand von Willebrand's factor. Defects in this last give rise to von Willebrand's disease, the most common congenital bleeding disorder. "Normal" platelet adhesion occurs in response to blood vessel damage. They also interact, however, with fractured atherosclerotic plaques forming deposits and aggregates that cause ischaemic damage in cardiovascular, cerebrovascular, and peripheral vascular disease.

Two types of epidermal blistering disease result from adhesive defects. Epidermolysis bullosa is caused by genetic defects in molecular components of the adhesion mechanisms that bind the epidermis to the dermis. Junctions called hemidesmosomes located on the basal surfaces of the basal keratinocytes provide structural links between the anchoring fibrils (composed of type VII collagen) in the dermis. The linkage is mediated by the α6β4 integrin of hemidesmosomes and the basement membrane components, laminin 5. A genetic defect in any one of these components results in blistering disease that may be fatal or result in severe morbidity (fig 8.4).

The sera of patients with the autoimmune blistering disease pemphigus contain antibodies to the desmosomal cadherins, desmogleins 1 or 3, whereas autoantibodies to a hemidesmosomal component, known as the bullous pemphigoid antigen 2 or collagen XVII, are present in sera from patients with bullous pemphigoid. These autoantibodies cause loss of cell adhesion by an as yet unknown mechanism.

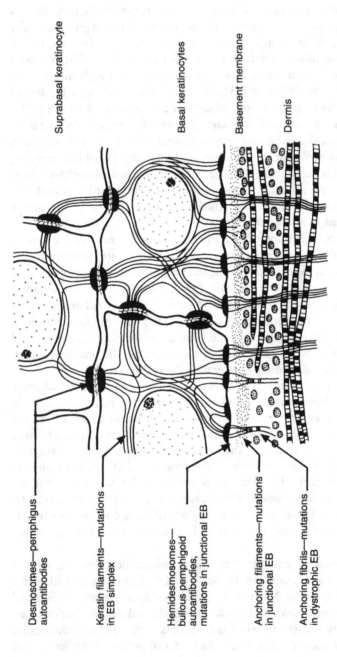

Suprabasal keratinocyte

Basal keratinocytes

Basement membrane

Dermis

Desmosomes—pemphigus autoantibodies

Keratin filaments—mutations in EB simplex

Hemidesmosomes—bullous pemphigoid autoantibodies, mutations in junctional EB

Anchoring filaments—mutations in junctional EB

Anchoring fibrils—mutations in dystrophic EB

FIG 8.4—*Basal epidermis and dermoepidermal junctions: diagram summarising how autoimmune and inherited diseases affect this region causing blistering diseases. (Reproduced from NATO Advanced Studies Institute Series, Volume 99, by permission of Plenum Press.)*

Defective cell adhesion has long been thought to play a part in the invasive and metastatic behaviour of neoplastic cells. Invasive cells spread into the tissues surrounding the primary tumour, penetrate into blood vessels, lymphatics, or body cavities, and become dispersed to distant areas. Some may become trapped at new sites, extravasate, and form secondary tumours. This is a complex series of events that may entail various altered cellular properties, such as secretion of proteolytic enzymes, cell motility, and growth, as well as changes in adhesiveness. A straightforward view of how adhesive changes might be involved in cancer is that reduced adhesion would promote the early events in invasion and metastasis when cells leave the primary site. Increased adhesion would, on the other hand, enhance the formation of metastatic deposits at remote sites. Both types of changes have been reported in various tumours. Mutations or reduced expression of E-cadherin, integrins, or desmosomal components has been found in several different tumour types, suggesting a possible tumour suppressor function for these molecules.[13] The tumour suppressor gene DCC (Deleted in Colorectal Cancer) was found by sequence analysis to code for a member of the immunoglobulin superfamily related to N-CAM, although whether this particular molecule actually functions in cell adhesion remains controversial. One variant form of the membrane proteoglycan, CD44, has been shown to promote metastasis in experimental systems and variants that may have metastasis promoting properties have been demonstrated in human cancer. Upregulation of adhesion molecules, for example, integrins in melanoma, may mediate secondary adhesion of metastatic tumour cells, thus promoting formation of secondary tumours.

Infections and parasitic diseases

Cell adhesion molecules also participate in some infectious and parasitic diseases, in which the invading organisms use the normal tissue molecules for binding. Thus the rhinoviruses—RNA viruses responsible for about half of common colds—bind to intercellular adhesion molecules on respiratory epithelium. Of more sinister importance, the binding of human immunodeficiency virus (HIV) during infection of T lymphocytes is mediated by binding of the viral gp120 protein to CD4, the immunoglobulin like T cell coreceptor. Intercellular adhesive molecules are also implicated in the adhesion of red blood cells infected with *Plasmodium falciparum* to capillary endothelium in the pathogenesis of severe malaria.[10] Cytokine mediated upregulation of endothelial intercellular adhesion molecules may be associated with severity of the disease.

Molecular approaches to diagnosis and therapy

The identification of single gene defects in components of adhesion systems enables prenatal diagnosis for diseases such as epidermolysis bullosa. Loss of function defects in single genes also offers potential for gene therapy. For example, it has been shown that adhesion can be restored to leucocytes from patients with LAD-I by transfection with an expression vector containing the cDNA for the normal β2 integrin subunit. Similarly, the invasiveness of an E-cadherin deficient mammary carcinoma cell line was suppressed by transfection with E-cadherin.[8]

Where disease involves increased adhesion various strategies for inhibition of adhesion are being investigated. An early example of this was the experimental inhibition of melanoma metastases in mice by injection of a synthetic peptide containing the –Arg–Gly–Asp– tripeptide that blocks integrin α5β1 binding to the matrix protein fibronectin.[6] Various peptides, carbohydrates, and monoclonal antibodies that have been shown to inhibit adhesion experimentally are being considered as possible therapeutic agents in inflammatory and clotting disorders.[11 12]

Other important functions of cell adhesion molecules

A complicating but also extremely exciting aspect of the study of cell adhesion is that, in addition to binding cells to each other or to the extracellular matrix, adhesion molecules transduce signals across cell membranes.[15] Such signals are involved in the regulation of almost every aspect of cellular function, including growth, movement, programmed cell death (apoptosis), and gene expression and differentiation. The interactions that adhesion molecules make with structural (cytoskeletal) and signalling molecules in the cell cytoplasm mediate these regulatory functions. This greatly increases the potential for involvement of adhesion mechanisms in disease.

1 Hynes RO. The complexity of platelet adhesion to extracellular matrices. *Thrombosis Haemostasis* 1991;66:40–3.

2 Edelman GM, Thiery JP, Cunningham BA, eds. *Morphoregulatory molecules.* New York: John Wiley, 1990.

3 Takeichi M. Cadherins: a molecular family important in selective cell-cell adhesion. *Annu Rev Biochem* 1990;59:237-52.

4 Legan PK, Collins JE, Garrod DR. The molecular biology of desmosomes and hemidesmosomes: "What's in a name?" *Bioessays* 1992;14:385-93.

5 Springer TA. Adhesion receptors of the immune system. *Nature* 1990;346: 425-34.

6 Humphries MJ, Olden K, Yamada KM. A synthetic peptide from fibronectin inhibits experimental metastasis of murine melanoma cells. *Science* 1986;223: 467-9.

7 Gould RJ, Polokoff MA, Friedman PA, Huang TF, Holt JC, Cook JJ, et al. Disintegrins: a family of integrin inhibiting proteins from viper venom. *Proc Soc Exp Biol Med* 1990;195:168-71.

8 Frixen UH, Behrens J, Sachs M, Eberle G, Voss B, Wards A, et al. E-cadherin-mediated cell-cell adhesion prevents invasiveness of human carcinoma cells. *J Cell Biol* 1991;113:173-85.

9 Fearon ER, Vogelstein B. A genetic model for colorectal tumorigenesis. *Cell* 1990;61:759-67.

10 Berendt AR, Simmons DC, Tansey J, Newbold CI, Marsh K. Intercellular adhesion molecule-1 in an endothelial cell adhesion receptor for *Plasmodium falciparum. Nature* 1989;341:57-9.

11 Ciba Foundation Symposium 189. *Cell adhesion and human disease.* Chichester: John Wiley & Sons, 1995.

12 Seiss W, Lorenz R, Weber PC, eds. *Adhesion molecules and cell signalling: biology and clinical applications. Topics in molecular medicine,* Vol 1. New York: Raven Press, 1995.

13 Hart I, Hogg N, eds. *Cell adhesion and cancer. Cancer surveys,* Vol 24. New York: Cold Spring Harbor Laboratory Press, 1995.

14 Collins JE, Garrod DR, Austin TX: eds. *The molecular biology of desmosomes and hemidesmosomes.* RG Landes Co.

15 Rosales C, O'Brien V, Korberg L, Juliano R. Signal transduction by cell adhesion receptors. *Biochim Biophys Acta* 1995;1242:77-98.

9: How do receptors at the cell surface transmit signals to the cell interior?

Robert H Michell

The chemical and geographical variety of biological signals

The cells of the body are bathed in fluids, the main constituents of which—for example, the ions and proteins of plasma—occur at almost constant concentrations. These extracellular fluids also contain, at very low and constantly changing concentrations, an incredible diversity of signalling molecules which, working together, control the behaviour of our cells and tissues and thus integrate body function (table 9.1). Surface molecules on cells also interact directly with partner molecules on adjacent cells to generate regulatory signals (table 9.1).

The most intimate signalling relationships are those that involve cell–cell contact, in which a receptor on one cell responds directly to ligation by a chemical structure displayed on the surface of its neighbour. At the other extreme, stimuli such as the classic hormones are widely and rapidly disseminated throughout the body, or at least throughout the extraneural tissues, but only initiate tissue specific responses in "target cells", that is, those cells that bear receptors responsive to them. Between these extremes of localisation and dissemination lie many extracellular signals, the actions of which affect larger volumes of tissue than classic neurotransmitters at close synapses, but are still substantially localised. For example, neurotransmitters and unstable locally generated compounds, such as nitric oxide (NO), are released from one cell to control others that are very close by, and peptide neurotransmitters may diffuse away from a nerve ending and act

TABLE 9.1—*Some examples of the great variety of stimuli to which mammalian cells can respond*

Light (vision)
Mechanical (for example, auditory receptors) or osmotic perturbation
Small ions (for example, Ca^{2+}) or inorganic compounds (for example, nitric oxide)
Smells and tastes, chemoattractants (for example, sexual)
Cell–cell interactions (Sertoli cells/developing sperm, sperm/egg, lymphocyte/antigen presenting cell, neutrophil/endothelium, etc)
Neurotransmitters (amines, amino acids, small peptides)
Hormones (small amines, small peptides, proteins)
Cytokines, colony stimulating factors, growth factors, and cell differentiation and survival regulators
Antigens: intact (B lymphocytes) or fragments arrayed on self-MHC (T lymphocytes)
Fatty acid metabolites (prostanoids, etc)
Inflammatory mediators (histamine, platelet activating factor, complement peptides, tachykinins, etc)
Extracellular matrix (ECM) components (fibronectin, laminin, collagen, etc) and/or growth factors bound to ECM

on cells a few tens of micrometres away before their destruction. Prostaglandins (prostacyclin from endothelium and thromboxane A_2 from activated platelets), subendothelial collagen, and adenosine diphosphate (ADP, secreted from platelets) regulate platelet and endothelial behaviour during incipient thrombus formation at sites of blood vessel injury; platelet derived and other polypeptide growth factors control the repair of these and other wounds locally. Histamine, leukotrienes, platelet activating factor (PAF), and lymphokines mediate local inflammatory events. Hypothalamic (neuro)hormones are carried the short distance to the anterior pituitary, their major target organ, by a specialised local circulation.

The control mechanisms used by cells have little regard for what types of molecules arrive as extracellular signals and whether information comes from nearby or from a distant source, so the nose's first reaction to an interesting smell and the response of an egg to contact with a fertilising sperm employ the same biochemical principles as the reactions of our major organs to hormones and neurotransmitters.

The heavy load imposed by signalling on the genome

Each cell in the body carries an identical genome, with the potential to encode about 80 000 proteins. How therefore do our diverse cells mount disciplined responses to such a plethora of external chemical influences? The research of the past few decades has provided a remarkably simple and elegant answer. Each extracellular stimulus "talks to" its own unique receptor protein (or, very often, to multiple types of receptor), but only a modest number of intracellular signal amplifying pathways are used to transmit the information from these multifarious receptors into cells. The first components of these signalling pathways "downstream" of the receptors are often built into the plasma membrane that encloses the responding cell.

In this way, the molecular components of one signalling pathway can be deployed to control many intracellular responses of diverse cells to miscellaneous extracellular signals. Sometimes, the intracellular signalling molecules are "second messengers" which change the activity of some pre-existing protein(s), and in other cases they are transcription factors which regulate the expression of gene(s) that encode proteins essential for the response. Key principles of this economical mode of transmission of information are summarised in figure 9.1.

Most extracellular stimuli are large and/or hydrophilic molecules which interact promptly and with high affinity with receptors on the cell surface. Notable exceptions to this general description include steroid hormones, thyroid hormones, and retinoids. Such lipid soluble molecules enter cells and bind directly to intracellular receptor proteins, with the resulting hormone–receptor complexes then interacting directly with chromatin to control gene expression. These lipophilic signals and their receptors will not be considered further in this chapter.

Recognition of extracellular stimuli

The exquisitely selective sites at the cell surface that recognise extracellular stimuli and often discriminate between closely related molecules (such as the nonapeptide neurohormones oxytocin and vasopressin) were named "receptors" long before we learned that they are membrane spanning proteins integrated into the structure of the plasma membrane. Those early studies often revealed striking heterogeneities in the quantitative pharmacological properties of different tissue responses to one stimulus, giving rise to the idea that a single extracellular signal molecule often activates more than one type of target cell receptor. Early examples of such multiple

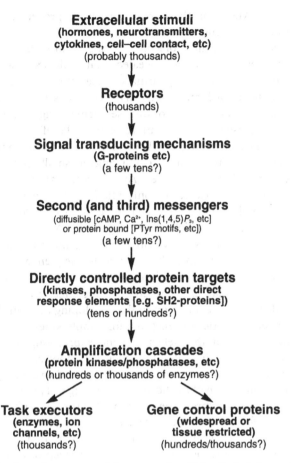

FIG 9.1—*A schematic summary of the manner in which cells feed the regulatory information brought by a very large number of stimuli, acting at an even larger number of receptors, to the cell interior through a much more limited number of signal transduction pathways.*

receptors—which were initially only elements of pharmacological hypotheses that aimed to explain unexpectedly complex experimental results—were the adrenoceptors of the α_1, α_2 and β types, muscarinic and nicotinic acetylcholine receptors, and H_1- and H_2-histamine receptors. Subsequently, it was recognised that many extracellular stimuli—including most or all amine neurotransmitters (table 9.1), small peptide ligands such as vasopressin, and larger peptides (for example, multiple members of the transforming growth factor β [TGF-β] family)—act at multiple receptors. Many of these receptors have proven or potential applications as drug targets.

In recent years, the molecular cloning of an enormous number of receptor proteins has revealed that cells express far more receptors than classical pharmacological studies ever indicated. The genome of each nucleated cell thus encodes even more receptor proteins than the number of diverse chemical signals, which have to be recognised by one or more of the cells of the body at one time or another during our lives. This amounts to maybe a few thousand receptor genes, occupying a small but significant percentage of the genome. At a particular time, any one differentiated cell expresses only relatively few of these receptor genes, and this enables the cell to respond only to that limited range of stimuli for which it is a "target cell". Moreover, each cell's receptor status is constantly being environmentally regulated, rather than static. For example, the complicated sequence of events that is initiated when antigens stimulate lymphocytes to proliferate and then to differentiate into mature immune effector cells is critically dependent on the expression in the developing cells, in the correct sequence, of a series of receptors for lymphokines.

The control of this selective gene expression, which is only now becoming understood, is immensely important. As an absurd illustration of its importance, imagine the endocrine mayhem that would ensue if each of the hormone secreting cell populations of the pituitary was to express the receptors for the wrong hypothalamic releasing factors: it would then secrete its stored hormones in response to the wrong hypothalamic stimuli. In the real world, there is growing recognition that incorrect receptor expression, for example, of serpentine receptors or receptor tyrosine kinases, is a relatively frequent contributor to human disease.

The transmission of receptor signals to the cell interior: general principles

The load on the genome

The heavy genetic load imposed by the cell's need to encode a huge variety of receptors would be further compounded if each receptor used different machinery to send its message onwards into the cell. This is not the case. As summarised in figure 9.1, evolution tends economically to employ one set of proteins to do multiple signalling jobs. The plasma membrane's biochemical machinery thus translates the body's extraordinarily complex language of extracellular stimuli, to much of which all cells are exposed, into a very much smaller lexicon of chemical, ionic, and electrical signals that control intracellular events. Figure 9.2 outlines some of the receptor controlled signalling mechanisms that are understood best and some of the receptors that use them (a more detailed version can be found in Michell[1]), and below a selection of the best understood and most important mechanisms is described in more detail. A recent book[2] and two single topic issues of *Trends in Biochemical Sciences*, on "signal transduction" (October 1992, volume 17, pages 367–443) and "protein phosphorylation" (November 1994, volume 19, pages 439–518), describe many of these very well. An annual suite of reviews in *Current Opinion in Cell Biology* regularly updates the continuing rapid advances.[3]

Receptor proteins that both recognise a stimulus and generate a signal

As depicted in figure 9.2, some receptor proteins incorporate in a single protein both an outward facing recognition site through which they respond to an extracellular stimulus and the mechanism that transmits the stimulus onwards to the cell interior. Receptors of this design include the large family of ligand gated ion channels (table 9.2) and a smaller group of ligand stimulated guanylate cyclases[4] (table 9.2). The receptor tyrosine kinases (RTKs) also include, in a single protein, both the external ligand recognition site and a cytoplasmically oriented catalytic domain that phosphorylates tyrosine (Tyr) residues both in the RTK molecule and on intracellular substrate proteins (table 9.2).

Multi-protein receptor–effector complexes

Many more cell surface receptors signal through multi-protein relay systems at the plasma membrane. Some hundreds (at least) of these receptors make up a single structural family of "serpentine" receptor proteins, with their ligand recognition sites exposed to the extracellular medium and their polypeptide chains traversing the plasma membrane seven times.[5][6] Such receptors respond to an enormous range of stimuli, including light, odours, amine neurotransmitters, protein hormones, the Ca^{2+} ion, small peptide

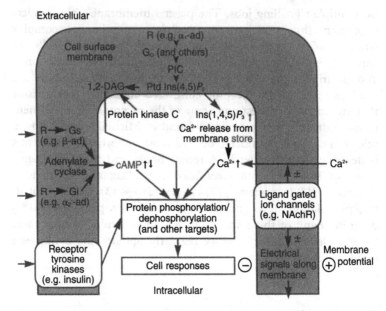

FIG 9.2—*Diagrammatic representation of some of the mechanisms of transmembrane signalling used by cell surface receptors. At the left are shown receptors possessing intrinsic protein tyrosine kinase activity, alongside the control of adenylate cyclase by stimulatory and inhibitory receptors (indicated as R) which is mediated by G_s and G_i. At the top is the receptor stimulated hydrolysis of PtdIns (4,5)P_2 (phosphatidylinositol 4,5-bisphosphate) by phosphoinositidase C (PIC), leading to intracellular accumulation of 1,2-diacylglycerol (1,2-DAG), Ins(1,4,5)P_3, and Ca^{2+}. At the right are the actions of receptors possessing intrinsic ion channels on membrane potential and hence on cellular electrical activity and Ca^{2+} permeability. Many, but not all, of the intracellular effects of changes in the cellular concentrations of cAMP, 1,2-DAG, and Ca^{2+} are caused by the phosphorylation and dephosphorylation of particular target proteins by the various kinases and phosphatases that these intracellular messengers regulate, as depicted in the centre.*

hormones, and the prenylated peptide mating factors of yeasts (table 9.2).[7][8] Each serpentine receptor communicates with one or more coupling proteins (G proteins) that are complexed to guanine nucleotide (GDP or GTP), and these control the activity of effector protein(s), usually either enzyme(s) or ion channel(s), that transmit the incoming signals to the cell interior[7][8] (figs 9.2 and table 9.2). The G proteins are heterotrimers of α, β, and γ subunits, with these subunits constituting three much smaller evolutionary families of related polypeptides—probably no more than 50 genes in all,

TABLE 9.2—*Some examples of the mechanisms of signal transmission used by receptors*

Receptors with intrinsic ion channels

Excitatory: acetylcholine (nicotinic receptors: primarily Na^+ selective channels)
glutamate ("NMDA receptor": relatively Ca^{2+} permeable)

Inhibitory: γ-aminobutyric acid (GABA$_A$) and glycine (Cl^- channels)

Receptors with intrinsic protein tyrosine kinase activity

Platelet-derived growth factor (PDGF)	Epidermal growth factor (EGF)
Insulin	Insulin-like growth factor 1 (IGF-1)
Macrophage colony stimulating factor (M-CSF)	Hepatocyte growth factor (HGF)
Haemopoietic stem cell factor (SCF/*Steel* factor)	Nerve growth factor (NGF)

Receptors that activate Jak family tyrosine kinases and Stats

Erythropoietin	Growth hormone
Ciliary neurotrophic factor (CNTF)	Granulocyte colony stimulating factor (GCSF)
Interleukins 2–7, 9, 10 and others	Thrombopoietin
Prolactin	Interferons

Receptors with intrinsic guanylate cyclase activity

Atrial natriuretic peptide	*E. coli* enterotoxin receptor (gut)

Serpentine receptors that activate adenylate cyclase

Adrenergic (β_1- and β_2-receptors)	Prostacyclin
Histamine (H$_2$-receptor)	Thyroid stimulating hormone (TSH)
Vasopressin (V$_2$ receptor)	Growth hormone releasing hormone (GHRH)
Adrenocorticotrophic hormone (ACTH)	Olfactory receptors (large family)
Glucagon	Some taste receptors?

Serpentine receptors that inhibit adenylate cyclase

Adrenergic (α_2-receptor)	Adenosine (A$_1$ receptor)
Prostaglandin E$_1$ (some tissues)	Acetylcholine (muscarinic—two subtypes)
Opioid peptides (δ receptor)	

Serpentine receptors that activate a cyclic nucleotide phosphodiesterase

Rhodopsin (vertebrates: four forms, in retinal rods and three types of colour selective cones)
Some taste receptors?

Serpentine receptors that open a hyperpolarising K^+ channel

Acetylcholine (muscarinic—one subtype)

Serpentine receptors that activate PtdIns(4,5)P$_2$ hydrolysis

Adrenergic (α_1-receptor)	Platelet activating factor (PAF or AGEPC)
Histamine (H$_1$-receptor)	Rhodopsin (invertebrates)
Acetylcholine (muscarinic; two subtypes)	Thromboxane A$_2$
Vasopressin (V$_1$ receptor)	Interleukin-8
Substance P/other tachykinins	Glutamate ("metabotropic" receptor)
Endothelins (three receptors)	Some taste receptors?

For references, see the text.

with the majority encoding α subunits. The effector proteins that the G proteins control and their downstream targets are, however, structurally and functionally much more diverse (table 9.2). Information is transmitted to the α subunits of G proteins mainly by the cytoplasmically oriented polypeptide loop between the fifth and sixth transmembrane domains of activated serpentine receptors.[5]

At least three major evolutionary outcomes have emerged from the versatility of the cooperation between the serpentine receptor protein superfamily and the three G protein subunit families. First, an enormous range of extracellular signals, from photons arriving at the retina to peptide hormones, transmit most or all of their cellular control information through a modest number of effector molecules. Second, multiple but related serpentine receptors for a single stimulus often employ different signalling mechanisms (for example, the various muscarinic acetylcholine receptor subtypes —see table 9.2). Third, closely related receptors sense and transmit the information that is delivered to cells by closely related extracellular signals (for example, oxytocin versus vasopressin, and the different neurokinins).

Receptors that control the formation of intracellular second messenger molecules

Adenylate cyclase and cAMP: the prototypic G protein mediated signalling pathway

In the 1950s, Earl Sutherland and his colleagues discovered that catecholamines (acting at β adrenoceptors), glucagon, and some other hormones transmit their messages to the cell interior by stimulating the synthesis by adenylate cyclase (also known as adenylyl cyclase) of adenosine cyclic 3':5'-monophosphate (cAMP), an intracellular nucleotide "second messenger" that is present in the cytosol at micromolar concentrations (see table 9.2). The resulting changes in the intracellular concentration of cAMP are sensed by one or both isozymes of a cAMP activated protein kinase: this kinase serves as an intracellular "receptor" for cAMP which, when activated, phosphorylates and changes the functions of many intracellular proteins, including key enzymes of central hormone regulated metabolic pathways such as muscle glycogen breakdown and adipose lipolysis.

When he made this discovery, Sutherland recognised that there would be multiple intracellular "second messengers" that would mediate hormone actions of which cAMP would be only the first, but he could hardly have guessed the degree to which the lessons learned from studies of cAMP formation and action would inform later studies of other signalling systems—and how long it would take to identify other second messengers. In particular, his laboratory's work, together with that of Edwin Krebs and Ed Fischer, established a principle central to most later discoveries in cell signalling research: *regulation of the functions of intracellular proteins, and thus the control of cell behaviour, is largely achieved by the reversible phosphorylation and dephosphorylation of pivotal regulatory proteins.* To this end, cell surface receptors for extracellular signals have evolved an astonishing armoury of different ways of controlling the activities of the kinases and phosphatases that add and remove phosphate groups from such proteins.

The receptors that exert control over cells by regulating adenylate cyclase activity—some stimulate cAMP formation and others inhibit—belong to the enormous superfamily of G protein coupled serpentine receptors which was introduced earlier. The extracellular information transmitted through the many adenylate cyclase coupled serpentine receptors is summarised for the cell interior primarily as a rise or fall in the intracellular concentration of cAMP (see fig 9.2, table 9.2; see also Pieroni *et al*[9] and Bentley and Beavo[10]). The first element of this simplification is demarcation of the receptors that regulate adenylate cyclase into stimulatory receptors and inhibitory receptors, with each subgroup communicating with adenylate cyclase through different heterotrimeric G proteins. The stimulatory and inhibitory G proteins are, respectively, G_s (two variants, derived from one gene: G_s is the cellular target of cholera toxin) and G_i (two distinct types, G_{i1} and G_{i2}, encoded by different genes: these are cellular targets for pertussis toxin).[8 9]

Receptor activation of heterotrimeric G proteins provokes their dissociation into free α subunits and $\beta\gamma$ complexes. During this dissociation, the guanine nucleotide binding site of G_α opens up, allowing it to exchange its bound guanine nucleotide with the cytosol. GTP is the most abundant cytosolic guanine nucleotide in healthy cells, so exchange generally replaces a bound GDP molecule, previously formed from GTP by the intrinsic GTPase activity of all G_α species, by GTP. As a result, the G_α subunits become ligated to GTP and functionally active.

Until recently, there was dispute over whether the relative quantities of stimulatory G_α species–GTP and inhibitory $G_{i\alpha}$–GTP freed by receptors into the membrane housing adenylate cyclase were solely responsible for the control of adenylate cyclase activity by receptors: other evidence suggested that liberated $\beta\gamma$ complexes might exert some of the inhibition caused by activated G_i. As is so often the case with scientific arguments that are initially framed as a choice between mutually contradictory scenarios, the answers that emerged combine elements of both views. Adenylate cyclase, like almost all other protein components of signalling pathways, exists in multiple isoenzymic forms (at least eight to date in mammals): GTP ligated G_α subunits are potent regulators of some adenylate cyclases, others are controlled by the local $\beta\gamma$ concentration, and yet others respond to the combined effects of G_α and $\beta\gamma$.[8 9]

Phosphatidylinositol 4,5-bisphosphate hydrolysis, a second ubiquitous pathway controlled through G proteins

A second large family of serpentine receptors feeds control cells through a very different signalling system based on the hydrolysis of the minor membrane phospholipid phosphatidylinositol 4,5-bisphosphate [PtdIns(4,5)P_2] by a specific phospholipase (phosphoinositidase C or PIC—see fig 9.2). In 1953, Hokin and Hokin observed that stimulation of receptors activates inositol phospholipid metabolism (see Michell[11]); two decades later the author realised that inositol lipid hydrolysis causes a rise in cytoplasmic Ca^{2+} concentration;[11] the early 1980s saw definition of how PtdIns(4,5)P_2 hydrolysis contributes to transmembrane signalling.[12]

As with adenylate cyclase, serpentine receptors activate G proteins that control multiple PIC isoenzymes (β_1, β_2, β_3). Coupling of PIC to many receptors (for example, the V_1 vasopressin receptor) is mediated by the G_Q subfamily of G proteins (G_Q, G_{11}, G_{16}), whose GTP ligated α subunits directly activate PIC-β_1: these G proteins are insensitive to cholera and pertussis toxins. However, pertussis toxin pre-treatment prevents PIC coupling to other serpentine receptors, including the receptor for the neutrophil chemotactic peptide fMet-Leu-Phe and some of those that inhibit adenylate cyclase. In these cases, the $\beta\gamma$ dimers liberated by G protein activation are the primary activators of PIC-β_2. PIC-β_3 can be activated both by $G_{Q\alpha}$ and by $\beta\gamma$.[8]

Hydrolysis of PtdIns(4,5)P$_2$ by PIC yields two products: water soluble inositol 1,4,5-trisphosphate [Ins(1,4,5)P$_3$] which diffuses into the cell interior, and lipid soluble 1,2-diacylglycerol (1,2-DAG) which remains associated with the cytoplasmic surface of the cell membrane.[12-17] Each has a unique role as an intracellular messenger. Within cells there are membrane compartments, particularly some regions of the smooth endoplasmic reticulum, into which Ca^{2+} is continuously sequestered by an ATP driven pump, so holding the cytoplasmic Ca^{2+} concentration of "resting" cells at about 0·1 µmol/l. When receptors provoke PIC activation and thus a rise in the intracellular concentration of Ins(1,4,5)P$_3$, this compound binds to receptors on the membrane around this Ca^{2+} store and triggers rapid Ca^{2+} release into the cytosol. The Ins(1,4,5)P$_3$ receptor proteins are a small family of oligomeric (probably tetrameric) ligand gated Ca^{2+} channels assembled from 250 000 dalton (Da) subunits: they probably function in a manner basically similar to the ligand-gated ion channel receptors of the plasma membrane (see below).[13 14]

As a result of stimulation, the cytoplasmic Ca^{2+} concentration often rises briefly to more than 0·5 µmol/l within seconds of the cell encountering a stimulus. Downstream of Ins(1,4,5)P$_3$-stimulated mobilisation of Ca^{2+}, cells often exhibit temporally and spatially complex changes in cytosolic Ca^{2+} concentration, notably travelling waves of [Ca^{2+}] that traverse the cell from an initiating site at an Ins(1,4,5)P$_3$-responsive Ca^{2+} store. These are most often seen as temporally separated spikes of whole cell [Ca^{2+}], the frequency of which is regulated by the agonist concentration.[13 14] Ins(1,4,5)P$_3$ is inactivated both by dephosphorylation and by entry into complex pathways that interconvert a number of previously unknown inositol polyphosphate isomers, the biological functions of which remain uncertain.[12 15]

The responses of cells to elevated [Ca^{2+}] are legion, including smooth muscle contraction, endocrine and exocrine secretion, control of metabolic pathways, and changes in the polarisation of excitable cells. Ca^{2+} controls some cellular targets directly, usually as a complex of Ca^{2+} and the Ca^{2+} binding protein calmodulin. For example, Ca^{2+}/calmodulin activates the Ca^{2+} pump of the plasma membrane, the activity of which is essential for cells effectively to terminate intracellular Ca^{2+} signals. In other cases, Ca^{2+} or Ca^{2+}–calmodulin controls the activity of protein kinase(s) and/or protein phosphatase(s), so changing the phosphorylation status and activity of intracellular targets, such as the myosin of smooth muscle.[16]

1,2-Diacylglycerol, the other messenger molecule formed from PtdIns(4,5)P_2, activates one or more of the protein kinases C (PKCs—a complex family of at least nine isozymic protein kinases), which can also be activated by phorbol ester tumour promoters.[17] These compounds were originally studied because they enhanced the tumour yield in skin treated with carcinogens. They were then found partially to mimic many effects of cell stimulation (such as platelet aggregation and lymphocyte proliferation), and the recognition that they activated PKCs confirmed that they are tumour promoters because they subvert a normal cellular signalling process.

The precise intracellular target molecules of the various PKCs are less well defined, but the cellular effects of permeating diacylglycerols and of tumour promoters that activate PKC are legion. Moreover, we do not yet fully understand either the mechanisms by which Ca^{2+} and PKC often act together in activating cells or by which they control longer term cellular responses such as expression of the interleukin-2 receptor and proliferation in lymphocytes.

Signalling through the opening of cell surface ion channels that are gated by ligands

The main permeability barrier within cell surface membranes is provided by the hydrocarbon side chains of membrane lipids, and the membranes normally have a very low permeability to small ions such as Na^+, Ca^{2+}, and Cl^-. Ion pumps driven by adenosine triphosphate (ATP) in the membranes constantly build ion gradients across this barrier, notably to keep the intracellular Na^+ concentration at least tenfold lower than the 100 mmol/l of extracellular fluid, and to keep the intracellular Ca^{2+} concentration ten thousandfold lower than the extracellular concentration (1–2 mmol/l).

These ion gradients (and others dependent on them, such as the excess of extracellular over intracellular Cl^-) make possible two types of signalling controlled by receptors. First, they give rise to a resting membrane potential (with the intracellular surface of the membrane electrically negative relative to the outside) which is perturbed by any selective and transient increases in membrane permeability to particular ions. Second, the low "resting" Ca^{2+} concentration in the cytoplasm (at about $0 \cdot 1$ μmol/l) means that small Ca^{2+} movements can substantially raise this value—as described above, a small rise in cytoplasmic Ca^{2+} can serve as an effective intracellular signal.

Nicotinic acetylcholine receptors[18-21]

The fast acting nicotinic acetylcholine receptor (NAChR) of skeletal muscles is the prototypic member of the ligand-gated ion channel receptor family. These multisubunit receptor proteins are transmembrane ion channels that open transiently when they bind agonists. The NAChR is strikingly abundant in the electric organs (electroplaques) of electric fishes (eels and rays). These evolutionary simplifications of skeletal muscles retain little other than the receptor-rich neuromuscular junctions of the progenitor muscles, so NAChRs of muscle type were the first receptors to be purified in quantity for detailed biochemical and structural study. They are variable five subunit proteins (typically $\alpha_2\beta\gamma\delta$ in fetal muscles and electroplaque; $\alpha_2\beta\epsilon\delta$ in adult muscle; molecular weight about 280 000) that span the cell surface membrane from outside to inside. Neuronal NAChRs are also pentamers, but in this case only combinations of α (eight or more subtypes) and non-α (also, confusingly, called β: three or more subtypes) subunits in varying stiochiometries.[20] The multiplicity of subunit isoforms, mostly encoded by different genes, allows diverse cells (fetal and adult muscles, central and peripheral neurons of various types, etc) to express multiple species of NAChR with different functional characteristics.

The individual NAChR subunits are all closely related and each folds so that a bundle of four membrane spanning sequences (M1–M4—about 20 residues long) crisscrosses the plasma membrane, with the N-terminus external and the C-terminus internal.[18 19 21] The five subunits form a quasi-symmetrical fivefold array around a central channel. One transmembrane α helix (M2) of each subunit is amphiphilic: one of its opposite faces "prefers" to associate with the hydrophobic core of the membrane and the other with the more polar environment of the ion channel. The five M2 helices contribute amino acid residues to a protected transmembrane pathway which allows cations through the membrane.[21] Na^+ is the preferred ion mobilised through most NAChRs by acetylcholine, but some $\alpha_n\beta_n$ pentamers are also fairly

Ca^{2+} permeable (that is, they resemble NMDA receptors—see below). Mutation of some of the amino acid residues that line the ion channel can greatly modify the ion gating kinetics and ion selectivity properties of a receptor, even to the extent of switching it between cation and anion permeability.[21]

Acetylcholine binding to ligand selective sites on the large extracellular domains of the α subunits provokes opening of the ion channel for a few milliseconds. Na^+ can then flow down its transmembrane gradient. The resting membrane potential of cells is normally negative inside (typically 50–80 mV—see fig 9.2), so cation influx reduces the membrane potential. The best known result of membrane depolarisation is the initiation of long distance electrical signals in the form of action potentials, as in nerve cells and at the neuromuscular junction. In some other cells, however, a major effect is to cause local changes in membrane permeability to key regulatory ions. For example, stimulation of NAChR causes catecholamine secretion from the adrenal medullary cells of some species: the membrane depolarisation triggered by receptors opens plasma membrane Ca^{2+} channels that are controlled by membrane potential, initiating a Ca^{2+} entry which triggers exocytosis of the hormone.

Excitatory receptors for glutamate[21] [22]

A second set of structurally similar receptors, including the so called NMDA receptors, mediates neuronal responses to excitatory amino acid neurotransmitters such as glutamate. These cation permeable receptors admit Ca^{2+} to cells fairly readily and are involved both in early stages of memory formation and in the neurodegenerative processes that are triggered if neurons are excessively and persistently excited.[22]

Inhibitory receptors

Receptors for inhibitory neurotransmitters (γ-aminobutyric acid [the $GABA_A$ receptors] and glycine) also include ligand-activated ion channels which belong to the same evolutionary superfamily as the excitatory receptors just discussed. However, negatively charged Cl^- ions flow into the responding neurons when these inhibitory receptor channels open in response to neurotransmitters:

the Cl$^-$ flux tends to hyperpolarise the neurons (raise their membrane potential), and this decreases their sensitivity to activation.

Receptors with intrinsic or associated protein tyrosine activity

Receptor tyrosine kinases

The receptor tyrosine kinases (RTKs) make up the largest and best understood family of receptor proteins that express an intrinsic enzyme activity. The first RTKs to be identified were the receptors for platelet-derived growth factor (PDGF) and epidermal growth factor (EGF) (see table 9.2): these were isolated as receptor proteins that catalyse the tyrosine phosphorylation both of their own cytoplasmic domains and of other proteins. When the EGF receptor was sequenced, it was realised that the previously cloned v-*erbB* oncogene encodes a constitutively active derivative of this receptor, and this immediately alerted researchers to the fact that aberrant RTK activity can contribute to carcinogenesis.

These initial discoveries were quickly followed by the identification—sometimes through direct cloning of oncogenes and sometimes by homology cloning, using oligonucleotides representing consensus regions from the first sequences determined —of many RTKs for which the biological ligands were at first unknown. Many of these "orphan" receptors have since had their functions defined. For example, the *met*, *trk*, *fms*, and *kit* proto-oncogenes encode, respectively, receptors for hepatocyte growth factor (HGF), nerve growth factor (NGF), macrophage colony stimulating factor (M-CSF) and haemopoietic stem cell factor (SCF).[23]

All RTKs incorporate both an extracellular ligand binding site (often for a growth factor) and a kinase domain at the cytoplasmic surface of the plasma membrane. In most RTKs, these two functional elements lie towards opposite ends of a single polypeptide chain which spans the plasma membrane. In apparent contrast, the insulin and insulin-like growth factor 1 (IGF-1) receptors are $\alpha_2\beta_2$ tetrameric proteins that are linked by disulphide bridges and include paired ligand binding sites (on each α) and paired kinase sites (on each β). These α and β subunits are initially made as a single polypeptide which is proteolytically processed before the migration of $\alpha_2\beta_2$ to the cell surface.

Stimulation of an RTK by its cognate ligand activates its tyrosine kinase activity, which initially catalyses tyrosine phosphorylation in the RTK's cytoplasmic domain. This is achieved by interchain transphosphorylation in a receptor dimer: the catalytic site of one chain phosphorylates tyrosine residues in the other. In the insulin and IGF-1 receptors, the hormone triggers transphosphorylation within a pre-existing receptor oligomer.[23] In other receptors, ligand binding drives receptor dimerisation and so sets the scene for transphosphorylation. Receptor phosphorylation has two main effects: it opens up the RTK's kinase site, converting it into a tyrosine kinase effective against other protein substrates, and it causes the cytoplasmic tail to display phosphotyrosine residues as signals.

The Ras/Raf/MAP kinase pathway

It has long been known that RTKs are only biologically functional if their tyrosine kinase domain is active, but only through a series of exciting recent discoveries have we learned why: figure 9.3 (route A) summarises the newly revealed signalling pathway. First came recognition that Ras (a small guanine nucleotide binding protein) and Raf (a serine/threonine directed protein kinase), two more proteins which, in constitutively active mutant forms, were first discovered as oncogenes, lie on the signalling pathway from RTKs to activation of cell proliferation. Then followed recognition that phosphotyrosine residues, when present in particular peptide sequence contexts either in the cytoplasmic domain of RTKs or in other substrate proteins, act as "tethered" second messengers. The motifs containing phosphotyrosine serve as high affinity binding sites to which intracellular signalling proteins bind through their SH2 (Src homology type 2) domains, and possibly also through a more recently recognised family of phosphotyrosine binding (PTB) domains.[24 25]

Different proteins containing SH2 domains associate with different motifs containing phosphotyrosine (PTyr): for example, P Tyr-Met-Asp/Pro-Met is recognised by the p85 regulatory subunit of phosphoinositide 3-kinase, PTyr-Ile-Tyr-Val by Src, and PTyr-Ile-Asn-Gln by Grb2. Through these interactions, activated RTKs and some of their substrate proteins—especially IRS-1, a multiply phosphorylated substrate of the insulin receptor—become

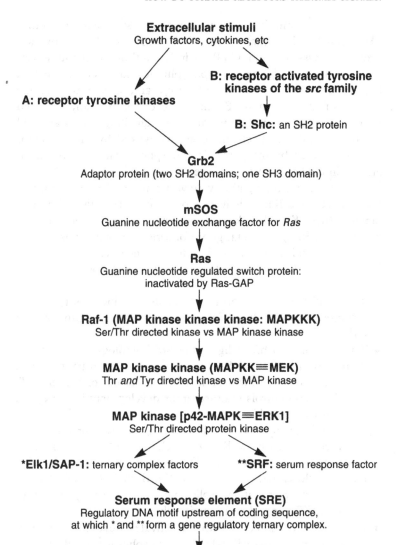

FIG 9.3—*The sequence of events that transmits signals from at least some RTKs (and other receptors that regulate non-covalently associated tyrosine kinases of the Src family). There are also routes to activation of MAP kinase by other receptors (for example, some serpentine G protein-coupled receptors; and some receptors that activate Jak kinases), and Raf-1 can transmit signals to the nucleus by at least one pathway that bypasses MAP kinases.*

assembled into multicomponent complexes with these and other SH2 proteins; these include Ras-GAP, a protein whose activation promotes Ras-catalysed GTP hydrolysis, thus accelerating Ras inactivation; the γ isozymes of phosphoinositidase C (see below); Syp (a protein tyrosine phosphatase); and numerous tyrosine kinases of the *src* family. By this type of mechanism, activated RTKs can initiate multiple intracellular signalling events and cell responses, and the spectrum of events evoked by each receptor can be dictated by the particular phosphotyrosine motifs that it and its immediate substrate proteins display.[23-26]

The best understood pathway downstream of RTKs is that which leads via Ras and Raf to activation of the mitogen-activated protein kinases (MAPKs—also known as extracellular ligand-activated kinases, or ERKs), a pivotal group of serine/threonine kinases, and thence to control of gene expression (fig 9.3) and often to mitogenesis. In essence, the key events in this pathway are as follows.[25-28]

- The synthesis of appropriate phosphotyrosine motif(s) on a receptor causes binding of the SH2 containing adaptor protein Grb2.
- This provokes Grb2 to ligate the Src homology type 3 (SH3) domain of mSOS; mSOS is the mammalian equivalent of "son of sevenless", the downstream effector protein of "sevenless", an RTK that controls photoreceptor development in the eye of the fruit fly *Drosophila.*
- mSOS is a guanine nucleotide exchange catalyst for Ras. Once activated, it provokes Ras to exchange its bound guanine nucleotide. In a cell, this usually means replacement of a GDP, formed endogenously on Ras by GTP hydrolysis, by a GTP (GTP is the most abundant cytosolic guanine nucleotide). Ras thus becomes GTP ligated and functionally active.
- GTP ligated Ras interacts with and activates the serine/threonine-directed protein kinase Raf-1, which phosphorylates and activates a downstream kinase known as MEK (MAPK or ERK kinase).
- Uniquely among known protein kinases, MEK then doubly phosphorylates MAPK (p42 and/or p44 isoenzymes) on both a threonine and a tyrosine residue within the highly conserved amino acid sequence motif:
  Phe-Leu-PThr-Glu-PTyr-Val-Ala-Thr-Arg-Trp-Tyr-Arg-Ala-Pro-Glu
- The activated MAPK thus becomes capable of phosphorylating a variety of substrates, and initiating a variety of cell specific

responses to the original stimulus. Responses can include protection of cells against possible death by apoptosis (for example, IGF-1, NGF), differentiation (for example, NGF), proliferation (for example, PDGF), activation of arachidonate mobilisation through phospholipase A_2 activation, etc.

• Selective control of nuclear gene expression is a key contributor to many of these responses. Figure 9.3 illustrates one such outcome: MAPK stimulates phosphorylation of transcription factor components such as SRF and the ternary complex factors Elk1 and SAP-1, leading to assembly of a gene-activating protein complex on the regulatory DNA sequence SRE.

In the past decade, definition of the RTK/Ras/MAPK pathway has been one of the triumphant vindications of the unitary biological philosophy which assumes that key signalling principles evolved in early eukaryotic (nucleated) cells and are still employed by all eukaryotes, and that these principles can be revealed by comparative analysis of the signalling pathways used by diverse organisms, including fission and budding yeasts (*Shizosaccharomyces pombe* and *Saccharomyces cerevisiae*), the fruit fly *Drosophila*, the nematode *Caenorhabditis*, and mammals (including transgenic mice and humans). The comment above on mSOS is one illustration of the stunning success of this approach in recent years.

A second payoff of this approach, initially in yeast and now emerging in mammals, has been recognition that most or all eukaryote cells house multiple signalling cascades of the general MAPKKK>MAPKK>MAPK>gene regulation design which is outlined in figure 9.3: *S. cerevisiae* has at least five such pathways, and mammalian cells are unlikely to be less complex. Each pathway is activated by a different set of environmental stimuli, and each pathway activates different, but often overlapping, sets of cell responses. For example, several cytokines (such as interleukin-1) and bacterial lipopolysaccharide can activate arachidonate mobilisation and prostanoid synthesis by activating p54, a MAPK like kinase known as stress-activated protein kinase (SAPK) (or Jun kinase-1, JNK1); its phosphorylation motif is PThr-Pro-PTyr in a sequence context slightly different from that in the canonical MAPKs. The SAPK pathway can probably also be activated via guanine nucleotide-binding protein/protein kinase partnerships other than that between Ras and Raf: candidates are Rac1 (and its relative Cdc42Hs) and the kinase PAK (p21 activated kinase).[28-30] The genes activated by the SAPK pathway include the transcription factors Jun and ATF2. Interleukin-1 and stimulators of the lymphocyte cell survival receptor CD40 activate the "p38 MAPK" pathway; this enzyme is a mammalian homologue of Hog-1, the terminal MAPK like kinase of a yeast osmotic sensing pathway (phosphorylation motif PThr-Gly-PTyr).

Activation of receptor associated tyrosine kinases

Transmembrane RTKs are not the only cell surface receptors that employ the activation of tyrosine kinase(s), leading to stimulation of the Ras/Raf/MAPKK/MAPK cascade and of other pathways triggered by phosphotyrosine motifs, to initiate their biological actions. Of other receptors that utilise tyrosine kinases, the best understood include:

- the multisubunit B lymphocyte immunoglobulin receptors, assembled from multiple Ig–α/Ig–β heterodimers
- the T lymphocyte antigen receptor assembly (a complex comprising α, β, and two ζ antigen receptor chains, together with three CD3 polypeptides (γδε)).[31]

On activation, these receptor complexes interact with and activate tyrosine kinases of the Src family (Lyn and/or Fyn in B cells; Lck and/or Fyn in T cells). Autophosphorylation opens up the catalytic sites of these kinases, which then phosphorylate immune cell-specific phosphotyrosine motifs (ITAMS), for instance on the ε and ζ chains of the T cell receptor array. The phosphorylated ITAMS then bind and activate Syk and/or ZAP-70, two related tyrosine kinases that include a pair of SH2 domains and transmit the activation message to the cell interior. One or more of these activated kinases phosphorylate Shc, an SH2 containing protein, and a phosphotyrosine motif on Shc signals Grb2 to activate mSOS, initiating the Ras/Raf-1/MAPKK/MAPK signalling cascade.

Jaks and Stats: a shorter tyrosine kinase-dependent route to gene activation[32][33]

Tyrosine phosphorylation has long been implicated in the actions of prolactin and growth hormone, and of many cytokines and haemopoietic factors, but their cloned receptors lack tyrosine kinase domains. Some of these stimuli, such as growth hormone, harness a unique receptor, but the dimeric receptors for a number of these agents share a common β subunit. This is a glycoprotein of molecular mass 130 kDa known as gp130 (or, sometimes, the closely related LIF receptor β subunit), and each receptor's ligand specificity is dictated by a committed α subunit.

As with the RTKs, ligand binding provokes these receptors to aggregate in the plane of the membrane. The oligomerised receptor complex then recruits tyrosine kinases of the Janus kinase (Jak) family: table 9.2 lists some of the ligands that employ Jaks for signalling. Three Jaks (Jak1, Jak2, and Tyk2) are widespread, but Jak 3 is found primarily in haemopoietic cells. Jaks bind to membrane proximal peptide sequences in the cytoplasmic domains of the receptors, and transphosphorylation between the Jaks that are brought together in the complexes unmasks their catalytic activity towards other proteins, particularly the Stat family of transcription factors (see below). Frequently, Jaks also phosphorylate the activated receptors, forming SH2 ligating phosphotyrosine motifs that activate the MAPK pathway, other events regulated by phosphotyrosine motifs (such as inositol lipid 3-kinase), and mitogenesis in a manner similar to RTKs (see above). Some such ligands, such as interleukin-4, harness the insulin receptor substrate IRS-1 to their cause.

The most characteristic action of the activated Jaks is rapid phosphorylation of proteins of the Stat family (signal transducers and activators of transcription), of which there are seven or more types. This provides a very economical route to nuclear gene regulation, but probably not to mitogenesis. Stats are transcriptional regulators of molecular mass 80–110 kDa; they all include an SH3 domain, an SH2 domain, and a tyrosine residue just C terminal the SH2 that can be phosphorylated. Once phosphorylated on this tyrosine, the initially inactive and cytosolic Stats pair into homodimers or heterodimers by mutual phosphotyrosine:SH2 coupling (they sometimes also associate with other protein partners). They then migrate into the nucleus, where they activate genes bearing appropriate regulatory sequences. For example, gene control and growth inhibition by interferon-γ primarily employs Stat1 and Stat2; stimulation of acute phase protein synthesis interleukin-6 by involves Stat3; prolactin harnesses Stat5 to control the mammary synthesis of some milk proteins; and interleukin-4 uses Stat6 to control immune function (immunoglobulin class switching, MHC class II expression, etc).

The future

Past research has identified, and given us a partial understanding of, some of the biochemical mechanisms by which cells respond to extracellular controls. Several are briefly discussed above, other established mechanisms and some receptors that employ them are mentioned in table 9.2, and yet others are just emerging or remain to be discovered.

In one sense, the picture so far is of great biological unity: many organisms control a panoply of events through a modest number of signalling mechanisms. On the other hand, evolution has harnessed a great variety of chemistries to this task—protein phosphorylation, lipid hydrolysis, ion flows, synthesis of NO—so some of the mechanisms that remain undiscovered may be quite similar to those we already know, whereas others will be total surprises. One major puzzle at present is how some serpentine receptors talk to the MAPKKK/MAPKK/MAPK cascade, and the emerging answer seems likely to involve the Rho subfamily of small G proteins.[29 30] Another puzzle is how, and to what end, many receptors control phospholipase D,[34] and here again a small G protein, probably Arf, seems to an important regulator in the pathway.[35]

As discussed above, we are beginning to understand how intracellular signalling pathways control complex cell behaviours such as the selective gene expression that is essential to successful cell differentiation. In the next few years we will see continuing major advances in this area, and should begin to understand much better the complex "cross talk" between signalling pathways that goes on inside cells.

1 Michell RH. Centrefold on transmembrane signalling. *Trends Pharmacol Sci* 1988;9 April issue
2 Barritt GJ. *Communication within animal cells.* Oxford: Oxford University Press, 1992.
3 *Current Opinion in Cell Biology*, review issues on cell regulation 1992;4:141–273, 1993;5:239–91, 1994;6:161–279, 1995;7:145–238,1996;8:137–244.
4 Thompson DK, Garbers DL. Guanylyl cyclase in cell signalling. *Curr Opin Cell Biol* 1990;2:206–11.
5 Coughlin SR. Expanding horizons for receptors coupled to G proteins: diversity and disease. *Curr Opin Cell Biol* 1994;6:191–7.
6 Baldwin JM. Structure and function of receptors coupled to G proteins. *Curr Opin Cell Biol* 1994;6:180–90.
7 Bourne HR, Sanders DA, McCormick F. The GTPase superfamily: conserved structure and molecular mechanism. *Nature* 1991;349:117–27.
8 Sternweis PC. The active role of βγ in signal transduction. *Curr Opin Cell Biol* 1994;6:198–203.
9 Pieroni JP, Jacobowitz O, Chen J, Iyengar R. Signal recognition and integration by G$_s$-stimulated adenylyl cyclases. *Curr Opin Neurobiol* 1993;3:345–51.
10 Bentley JK, Beavo JA. Regulation and function of cyclic nucleotides. *Curr Opin Cell Biol* 1992;4:233–40.

11 Michell RH. Inositol phospholipids and cell surface receptor function. *Biochim Biophys Acta* 1976;415:81–147.
12 Berridge MJ, Irvine RF. Inositol phosphates and cell signalling. *Nature* 1989; 341:197–205.
13 Tsien RW, Tsien RY. Calcium channels, stores and oscillations. *Annu Rev Cell Biol* 1990;6:715–60.
14 Miyazaki S. Inositol trisphosphate receptor mediated spatiotemporal calcium signalling. *Curr Opin Cell Biol* 1995;7:90–196.
15 Hughes PJ, Michell RH. Novel inositol-containing phospholipids and phosphates: their synthesis and possible new roles in cellular signalling. *Curr Opin Neurobiol* 1993;3:383–400.
16 Schulman, H. The multifunctional Ca^{2+}/calmodulin-dependent protein kinases. *Curr Opin Cell Biol* 1993;5:247–53.
17 Nishizuka Y. The molecular heterogenity of protein kinase C and its implications for cellular regulation. *Nature* 1990;334:661–5.
18 Unwin N. The nature of ion channels in membranes of excitable cells. *Neuron* 1989;3:665–76.
19 Karlin A. Structure of nicotinic acetylcholine receptors. *Curr Opin Neurobiol* 1993;3:299–309.
20 Role LW. Diversity in primary structure and function of neuronal nicotinic acetylcholine receptors. *Curr Opin Neurobiol* 1992;2:254–62.
21 Bertrand D, Galzi J-L, Devilliers-Thiéry, Bertrand S, Changeux JP. Stratification of the channel domain in neurotransmitter receptors. *Curr Opin Cell Biol* 1993; 5:688–93.
22 Collingridge GC, Singer W. Excitatory aminoacid receptors and synaptic plasticity. *Trends Pharmacol Sci* 1990;11:290–7
23 Lemmon MA, Schlessinger J. Regulation of signal transduction and signal diversity by receptor oligomerization. *Trends Biochem Sci* 1994;19:459–63.
24 Pawson T, Schlessinger J. SH2 and SH3 domains. *Curr Biol* 1993;3:434–43.
25 Songyang Z, Cantley LC. Regulation and specificity in protein tyrosine kinase-mediated signalling. *Trends Biochem Sci* 1995;20:470–5.
26 White MF. The IRS-signalling system in insulin and cytokine action. *Phil Trans R Soc Lond B* 1996;351:181–9.
27 Johnson G, Vaillancourt RR. Sequential protein kinase reactions controlling cell growth and differentiation. *Curr Opin Cell Biol* 1994;6:230–8.
28 Waskiewitz AJ, Cooper JA. Mitogen and stress response pathways: MAP kinase and phosphatase regulations in mammals and yeast. *Curr Opin Cell Biol* 1995; 7:798–805.
29 Symons M. Rho family GTPases: the cytoskeleton and beyond. *Trends Biochem Sci* 1996;21:178–81.
30 Malarkev K, Belham CM, Paul A, Graham A, McLees A, Scott PH, et al. The regulation of tyrosine kinase pathways by growth factor and G protein-coupled receptors. *Biochem J* 1995;309:361–70.
31 DeFranco AL. Transmembrane signalling by antigen receptors of B and T lymphocytes. *Curr Opin Cell Biol* 1995;7:163–75.
32 Ihle JW. Cytokine receptor signalling. *Nature* 1994;377:591–4.
33 Wells JA. Structural and functional basis for hormone binding and receptor oligomerization. *Curr Opin Cell Biol* 1994;6:163–73.
34 Billah MM, Anthes JC. The regulation and cellular function of phosphatidylcholine hydrolysis. *Biochem J* 1990;269:281–91.
35 Martin A, Brown F, Hodgkin MN, Bradwell AJ, Cook C, Hart M, et al. Activation of phospholipase D and phosphatidylinositol 4-phosphate 5-kinase in HL60 membranes is mediated by endogenous Arf but not Rho. *J Biol Chem* 1996; in press.

10: Membrane traffic, from cell to clinic

John Armstrong

Introduction: compartments and topology

The living cell is partitioned into different organelles, or compartments, each of which is specialised to carry out different functions. To do this the different compartments must be equipped with distinct complements of enzymes and other functional proteins. Yet nearly all the cell's proteins are encoded by genes that reside in the nucleus, and that are transcribed into RNA, then translated into protein, by the same cellular machinery in each case. It follows that the information specifying a protein's destination in the cell is in general included in its sequence of amino acids. How does the cell interpret this information and move proteins to different places? This chapter outlines some of what is known of the answers to this question, and uses, for illustration, examples of direct clinical relevance.

The first sorting decision is made before synthesis of the protein has finished. The *N*-terminal sequence of many proteins, the first part to be synthesised and emerge from the ribosome, forms a "signal" which is recognised by a particle in the cytoplasm. The complex of particle, nascent protein, ribosome, and RNA then docks with the surface of the endoplasmic reticulum (ER). As translation continues, the protein is extruded across the ER's membrane (fig 10.1). If the protein contains regions of particularly hydrophobic amino acids, however, the "translocation" process can be halted and the rest of the protein is left outside the ER. Thus the amino acid sequence of the protein determines its topology relative to the ER: either soluble within it or embedded in its

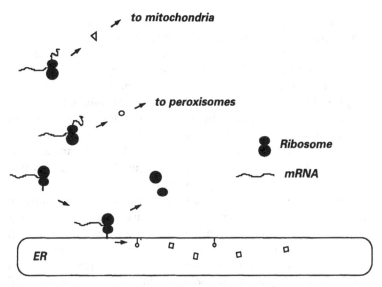

FIG 10.1—*The first stages in protein sorting. The first part of the mRNA for secreted and membrane proteins encodes a "signal sequence" which targets the translating complex to the ER membrane, where synthesis of the rest of the protein takes place. In contrast, mitochondrial and peroxisomal proteins are made in the cytoplasm, then transported to their destination organelles.*

membrane. Once this topology is established, it does not generally change in the protein's lifetime.

However, this "signal sequence" pathway operates not only for ER proteins, but also for proteins destined for all of the organelles of secretion and endocytosis: the Golgi complex, the plasma membrane and the lysosomes, and for secretion out of the cell. But the ER, and the other organelles, are closed and physically separated compartments. How, then, are proteins transported between these compartments?

Leaving the ER: quality control and cystic fibrosis

Traffic between different membrane compartments is by way of vesicles—patches of membrane that bend "outwards" from the organelle (that is, into the cytoplasm) and then pinch off completely, taking with them proteins from both the organelle's membrane and its soluble contents. The vesicles then travel through the cytoplasm to whichever is their destination membrane, and fuse with it. Both the membrane bound and soluble proteins are now components of the destination organelle, and their topology relative to the membrane and the cytoplasm is unchanged.

The first step of vesicle traffic, then, is from the ER to the Golgi complex. The vesicles, however, do not simply sample at random the contents of the ER. The ER is a site for more than just translation of proteins: the proteins must fold, assemble, and acquire various additional modifications before they are allowed out and into the secretory pathway. Proteins that fail this "quality control test" are often retained in the ER.

This test appears to be the cause of the most common form of cystic fibrosis. The defective gene in cystic fibrosis patients normally encodes the cystic fibrosis transmembrane conductance regulator, a plasma membrane protein involved in transport of chloride and other ions across the membrane. The mutant gene encodes a protein lacking a single amino acid. Surprisingly, the mutated protein seems to function almost normally as an ion transporter. Why, then, does it cause such drastic symptoms? The answer is that it fails the quality control test: it is trapped in the ER. But the failure is by a remarkably subtle margin—a slight drop in temperature allows the protein to assemble correctly and be transported to the plasma membrane.[1]

Retention in the ER: a novel approach to HIV immunisation

As the ER carries out this set of cellular functions, it must possess a specific set of functional proteins. Why do these proteins stay in the ER and not leave? For this, another signal is required. The soluble ER proteins have a specific amino acid sequence at the C terminus, KDEL in the one-letter code, which is recognised by a membrane receptor protein. In fact the "KDEL" proteins can leave the ER, but in the Golgi complex they attach to the receptor and are selectively retrieved back to the ER.

An understanding of this process makes it possible to redirect proteins within the cell. An example of this is a current approach to blocking the spread of infection by HIV (fig 10.2).[2] Like many "enveloped" viruses, HIV has a membrane glycoprotein which is made in the ER and then transported to the plasma membrane. Virus capsids containing nucleic acid can then attach, causing the

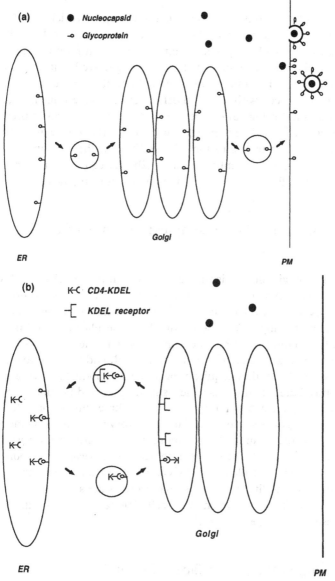

FIG 10.2—*Blocking the spread of HIV.*[2] *Normally (a) the HIV glycoprotein is made in the ER and travels by vesicles to the Golgi complex then the plasma membrane. There, the viral nucleocapsid attaches to it and the viruses bud out of the cell. When the cell expresses soluble CD4 modified with a KDEL signal (b), the glycoprotein attaches to CD4 in the ER instead of on the surface of uninfected cells. The KDEL signal then prevents the glycoprotein from reaching the plasma membrane, and hence prevents viral budding.*

virus to "bud" out of the cell, like a transport vesicle but in the opposite direction (fig 10.2a). The free virus infects other cells by attaching to CD4, a plasma membrane protein found in a subset of T cells. The gene for CD4 has been engineered to interfere with this process. Its transmembrane hydrophobic amino acids have been replaced with the KDEL signal, converting a plasma membrane protein into a resident of the ER. If cells expressing this altered CD4 are infected with HIV, the newly made viral glycoprotein attaches to the CD4–KDEL in the ER, and cannot reach the plasma membrane (fig 10.2b). Thus new viruses cannot bud out of the cell, and the infection cannot propagate.

Sorting in the Golgi complex: the basis of I cell disease

The Golgi complex is the next station on the secretory pathway. As well as carrying out a further series of modifications to proteins as they pass through, the Golgi complex has a sorting function: proteins leaving it may be destined for the plasma membrane or the lysosomes (fig 10.3). The degradative enzymes of the lysosomes are made, like secreted proteins, in the ER, and travel to the Golgi complex. There they acquire a specific sugar modification, mannose 6-phosphate. In this case the "signal" for modification is not a linear sequence of amino acids, but a three dimensional feature that distinguishes lysosomal from secretory proteins. A membrane receptor then binds the mannose 6-phosphate and diverts the lysosomal proteins into a separate pathway of vesicle transport out of the Golgi complex. If cells cannot carry out the mannose 6-phosphate modification, lysosomal proteins are secreted out of the cell. This is what happens in I cell disease, one of the inherited lysosomal storage diseases.[3]

Sorting at the cell surface: LDL and hypercholesterolaemia

As well as delivering material to the cell surface, cells require a second pathway of vesicle traffic to take up materials from the outside. This is called endocytosis, and its endpoint is the lysosome. Mechanistically, this route has much in common with secretion from the ER to the cell surface; however, its first step can be quite

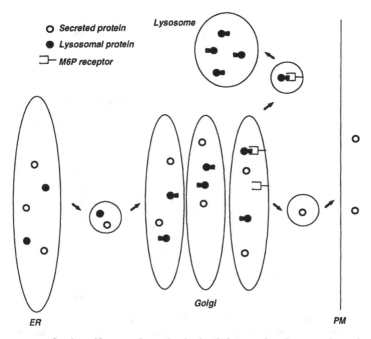

FIG 10.3—*Sorting of lysosomal proteins in the Golgi complex. Lysosomal proteins are recognised and modified with mannose 6-phosphate. The M6P receptor then diverts lysosomal proteins out of the secretory pathway and to the lysosomes.*

selective. Membrane proteins must first localise to specialised regions of the plasma membrane called coated pits, because it is these that "bud" inwards to form endocytic vesicles. This again requires a sorting signal.

An example of a protein possessing such a signal is the receptor for low density lipoprotein (LDL) (fig 10.4). Plasma cholesterol associates with LDL, which binds to the receptor and is taken up by endocytosis and delivered to the lysosomes, where the cholesterol is metabolised. Thus, the failure of any aspect of this process leads to hypercholesterolaemia. In one example of this condition,[4] the patient was found to possess normal amounts of LDL receptors which were quite capable of binding LDL. The receptors were, however, randomly spread across the plasma membrane instead of concentrating in coated pits; hence cholesterol was not removed and metabolised. Molecular methods revealed that this mutant receptor is altered in only a single amino acid in the short region

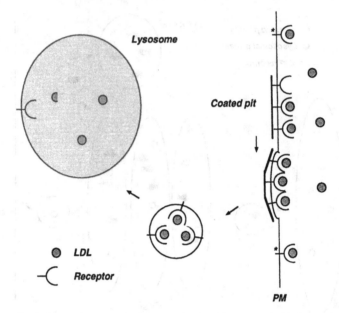

FIG 10.4—*Different strategies for endocytosis: the LDL receptor. Cholesterol binds to serum low density lipoprotein which attaches to LDL receptors in coated pits. These form vesicles which transport the LDL and receptor to the lysosomes, where cholesterol is liberated. Mutant LDL receptor (starred) lacks the signal to concentrate in coated pits, and thus does not enter the cell.*

of the receptor that extends into the cytoplasm. This amino acid is, indeed, critical to form the signal that allows selected membrane proteins to cluster in coated pits.

Regulated endocytosis: receptors for EGF and transferrin

The endocytic pathway has the capacity for more sophisticated options than continuous uptake and degradation in lysosomes. One example is the receptor for epidermal growth factor (EGF), which has the job of staying on the surface until EGF arrives from the serum (fig 10.5). Binding of EGF triggers a cascade of signals originating at the receptor, leading to the cell's elaborate responses to the growth factor. In the continuous presence of EGF, however, the cell might dangerously overreact. Thus it has developed the capacity to "downregulate". One consequence of EGF binding is

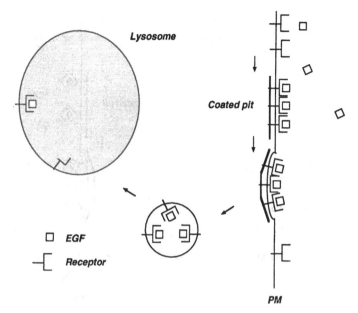

FIG 10.5—*Different strategies for endocytosis: the EGF receptor. Unbound receptor is not concentrated in coated pits. Binding of epidermal growth factor triggers the formation of a signal for the receptor to accumulate in coated pits and be "downregulated" by endocytosis.*

that the receptor migrates into the coated pits, from where endocytosis delivers it to lysosomes for degradation. Thus, once the signal has been sent, the signalling molecule is selectively removed.

A more subtle use of the endocytic pathway is involved in uptake of iron (fig 10.6). This uses two proteins, the iron binding transferrin and its membrane receptor, and cleverly exploits their sensitivity to both bound iron and acidity. In the serum, iron ions bind tightly to transferrin, which is now able to bind to the receptor. The complex of receptor, transferrin, and iron enters the endocytic pathway. As soon as endocytic vesicles leave the cell surface, a process of acidification begins. Lowered pH causes the iron to be released from transferrin, allowing its uptake across the vesicle membrane and into the cytoplasm. Rather than waste the two proteins, however, the transferrin receptor and bound transferrin are sorted in the endocytic pathway into vesicles which recycle it

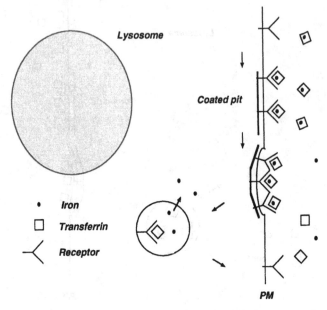

FIG 10.6—*Different strategies for endocytosis: the transferrin receptor. At serum pH, iron binds to transferrin, and this, but not free transferrin, binds to the receptor. In the endocytic pathway conditions become more acidic, causing iron to be released from transferrin. It is then transported across the membrane into the cytoplasm. At this pH free transferrin remains bound to the receptor, which follows a recycling pathway back to the plasma membrane. Here the free transferrin is released, making it and the receptor available for reuse.*

to the surface. At this relatively neutral pH, free transferrin can then dissociate and both proteins are available for another cycle of uptake.

Targeting the vesicles: SNAREs and neurotoxins

Transport between the different organelles is mediated by different vesicles, each originating in one place and destined for another. But how do the different vesicles know where to go?

This process is not yet fully understood, but several types of proteins have been implicated. One group is sometimes called SNAREs: a SNARE protein on the outside of the vesicle (the cytoplasmic surface) recognises another SNARE on the target membrane. Each combination of vesicle and destination may be specified by a particular group of SNAREs. Human scientists identified these proteins only recently, but certain bacteria have known about them for some time.

Neurons contain particular types of vesicles which are packed with neurotransmitters. These vesicles accumulate below the plasma membrane and, in response to a nerve impulse, fuse with it, rapidly releasing the neurotransmitters into the synapse. Neurotoxins from *Clostridium botulinum* and related bacteria act by specifically cleaving and inactivating one of the SNAREs involved in fusion of neurotransmitter vesicles. Remarkably, each SNARE so far identified in this process is the target of a neurotoxin.[5]

Mitochondria and peroxisomes: hyperoxaluria and molecular schizophrenia

Some organelles, particularly the mitochondria and peroxisomes, do not form part of the secretory or endocytic system. A few mitochondrial proteins are encoded in its tiny genome and made within it; the rest, and all the peroxisomal proteins, are made in the cytoplasm, and possess specific signals allowing them to bind and translocate into their respective organelles (see fig 10.1).

Here, too, defects in the targeting process are of clinical importance. Perhaps the strangest example comes from a form of hyperoxaluria.[6] The enzyme alanine–glyoxalate aminotransferase functions in the peroxisomes of humans, but in the mitochondria of some other species. One mutant form has a defective peroxisomal targeting signal. The mutant protein, however, is not simply left adrift in the cytoplasm. Instead, the mutation uncovers a hidden targeting signal for mitochondria, to which the protein is delivered, and the ensuing confusion of metabolism results in disease.

Conclusion

In this brief outline of some of the processes of protein targeting, nothing has been said of the different approaches of basic science, from cell biology, biochemistry, and yeast genetics, which have led to our present state of knowledge, and which are being used to resolve the many questions which remain. Perhaps it will be obvious that this basic science has already allowed an understanding of the pathology of a variety of human diseases, and has opened the way to entirely new sorts of therapy.

1 Denning GM, Anderson MP, Amata JF, Marshall J, Smith AE, Welsh MJ. Processing of mutant cystic-fibrosis transmembrane conductance regulator is temperature-sensitive. *Nature* 1992;**358**:761–4.
2 Buonocore L, Rose JK. Blockade of human-immunodeficiency-virus type-1 production in CD4[+] T-cells by an intracellular CD4 expressed under control of the viral long terminal repeat. *Proc Natl Acad Sci USA* 1993;**90**:2695–9.
3 Kornfeld S, Mellman I. The biogenesis of lysosomes. *Annu Rev Cell Biol* 1989; **5**:483–525.
4 Davis CG, Lehrman MA, Russell DW, Anderson RGW, Brown M, Goldstein JL. The J.D. mutation in familial hypercholesterolemia: amino acid substitution in cytoplasmic domain impedes internalization of LDL receptors. *Cell* 1986;**45**: 15–24.
5 Niemann H, Blasi J, Jahn R. Clostridial neurotoxins: new tools for dissecting exocytosis. *Trends Cell Biol* 1994;**4**:179–85.
6 Purdue PE, Allsop J, Isaya G, Rosenberg LE, Danpure CJ. Mistargeting of peroxisomal L-alanine-glyoxalate aminotransferase to mitochondria in primary hyperoxaluria patients depends upon activation of a cryptic mitochondrial targeting sequence by a point mutation. *Proc Natl Acad Sci USA* 1991;**88**: 10900–4.

11: Cytoskeleton and disease

Durward Lawson

All eukaryotic cells undergo a variety of complex cell movements ranging from cell migration to engulfing pathogenic bacteria and cell division. This ability is conferred by a cytoskeleton of three protein filament systems, namely microtubules, intermediate filaments, and actin filaments. This cytoskeletal network is dynamic and actively responds to a multitude of signals from both the extracellular environment and within the cell itself. For example, during embryonic development cytoskeletal activation leads to the migration of cells from the developing zygote to form the placenta. At a later stage in development, neural crest cells migrate and form peripheral neurons and Schwann cells. Within the developing nervous system neurons crawl vast distances to reach their target organ, whilst the molecules that pass down axons are both guided and driven by motor proteins associated with the cytoskeleton. Thus the final developmental reorganisation of the organism is attained. Given the role of the cytoskeleton in development, it is not surprising that it is involved in the tissue repair that mimics development, for example, fibroblast and epithelial cell locomotion is essential for wound healing. Similarly, in the immune system cell shape changes occur when dendritic cells process and present antigens as a response to pathogenic stimulus.

In most cells, the cytoskeleton forms a complex, apparently random, three dimensional meshwork, although in some cells such as neurons it is organised into parallel arrays. These variations in organisation are known to be related to function because the parallel axonal microtubules present in neurons act as "rail tracks" for vesicle transport, whilst the crosslinked meshwork of actin filaments, present at the leading edge of the neuronal growth cone and all motile cells, is necessary for their movement. In non-muscle cells these complex molecular organisations are controlled by at least 150 cytoskeletal associated proteins, and it is now clear that molecular alterations to either the cytoskeleton or molecules associated with it can cause disease (in the case of Duchenne muscular dystrophy) or, at the very least, occur in parallel with disease, for example, Alzheimer's disease and the oncogenically activated migration of tumour cells through tissue that occurs during metastasis and the development of cancer.

Microtubules and microtubule associated proteins
(fig 11.1)

Microtubules are essential for cell shape, polarity, organelle transport, and mitosis. They are formed from α and β tubulin monomers and, although there are at least six different types of tubulin monomers present in cells, they can and do polymerise with each other in vitro to form microtubules. This ability is not restricted to microtubules, but can also be found in actin filaments (see below) and is very likely to reflect the high degree of conservation found throughout evolution in many cytoskeletal

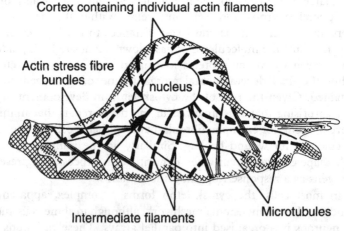

FIG 11.1—*Fibroblast attached to a flat substratum with the front of the cell at the left hand side. Actin is present as both individual filaments, in the outer cortical layer of the cytoplasm, and actin stress fibre bundles formed from many individual actin filaments which traverse the cytoplasm further inside the cell. Microtubules and intermediate filaments radiate from near the nucleus to the cell periphery.*

proteins. Microtubules are dynamic structures and they respond to the cell's requirements by depolymerising and polymerising using a pool of α and β tubulin subunits in conjunction with GTP hydrolysis. This is controlled by a feedback mechanism whereby the level of tubulin synthesis is directly regulated by a pool of soluble tubulin subunits the size of which is, in turn, dependent on the degree of microtubule assembly. In most animal cells, microtubules radiate out in all directions from a structure called the centrosome—the "control centre" for microtubule nucleation which interacts with the minus end of the microtubule, whereas subunit addition occurs at the plus end.

Many proteins are now known to associate with microtubules and these fall into two main classes: first, the microtubule associated proteins (MAPs) MAP 1a, 1b, 2a, 2b, 2, 4, and tau and, second, the motor proteins, kinesin and dynein, both of which use ATP hydrolysis as their energy source. MAPs are differentially expressed during development and their roles involve axonal microtubule stabilisation and the compartmentalisation of the nerve cell cytoplasm into axons and the nerve cell body, each with different receptors and ion channels. Tau is a neuronal phosphoprotein which is thought to stabilise axonal microtubules by binding to them, promoting their nucleation/elongation, protecting them from disassembly, and inducing them to form bundles. It therefore helps to provide a stable basis for axonal transport along the microtubule—a suggestion reinforced by the finding that, when this protein is transfected into non-neuronal cells, microtubules become more stable and the cells develop processes similar to neurites.

Microtubule associated motor proteins such as dynein act by driving the movement of intracellular structures as diverse as membranous organelles and chromosomes. The role of dynein has been particularly well defined in axonal transport, where it acts by transporting membrane vesicles away from the axon tip and towards the cell body and minus end of the microtubule, a process known as retrograde transport. Kinesin, on the other hand, transports membrane vesicles such as endoplasmic reticulum, lysosomes, pigment granules, and synaptic vesicles rapidly towards the axon tip and plus end of the microtubule, and away from the cell body—a process known as anterograde transport. Direct evidence for the role of kinesin has been shown in cultured neurons where inhibition of kinesin synthesis by antisense oligonucleotides results in a 10% reduction in axonal transport and a concomitant inhibition of neurite extension. In the nine years since "conventional" kinesin was first described, it is remarkable that at least 30 "kinesin related proteins" have been identified with multiple members of this genetic superfamily found in one organism.

Microtubules, microtubule associated proteins, and Alzheimer's disease

Alzheimer's disease is the fourth leading cause of death in the developed world after heart disease, cancer, and stroke. It is a disease that affects a high proportion of elderly people and, at present, there is no cure. Initial symptoms are problems in forming new memories but, as this form of senile dementia progresses, patients with Alzheimer's disease lose more and more higher functions. Although the pathology of Alzheimer's disease is complex, there are two major defining characteristics: first, extracellular neuritic plaques which contain deposits of β amyloid (a proteolytic fragment derived from a much larger precursor transmembrane glycoprotein) and, second, intracellular neurofibrillary tangles which consist largely of paired helical filaments (PHFs) formed from tau. The degree of dementia shows higher correlation with the extent of neurofibrillary pathology than with the number of extracellular neuritic plaques. During the course of Alzheimer's disease, tau becomes hyperphosphorylated at multiple sites, can no longer bind to and stabilise microtubules, and redistributes to the nerve cell body and dendrites where it assembles into PHFs. Direct evidence for the involvement of tau in Alzheimer's disease comes from studies in which transgenic mice overexpressed the largest human tau isoform. In these animals, human tau was aberrantly phosphorylated and mislocalised to nerve cell bodies and dendrites, as it does in humans with Alzheimer's disease. Furthermore, transgenic tau was found in association with the pre-tangle changes that precede the full neurofibrillary tangle pathology associated with Alzheimer's disease. These data do not, of course, tell us why tau mislocalisation is important in Alzheimer's disease but it is possible that it affects microtubule stability, disrupts axonal transport, and thus, over a period of many years, causes the neurodegeneration that leads, eventually, to Alzheimer's disease.

Intermediate filaments

Intermediate filaments (IFs) are found in the cytoplasm of most multicellular eukaryotes, but they have not as yet been identified in unicellular eukaryotes. Several different types of intermediate filaments have been described and all have a tissue specific distribution as follows: keratin (epithelia), vimentin (mesenchyme), desmin (muscle), glial fibrillar acidic protein (GFAP) and peripherin (peripheral neurons), neurofilaments and α internexin (CNS neurons), nestin (neuroectodermal stem cells), and lamins (nucleus). The progenitor of these apparently disparate proteins is likely to be a nuclear lamin with gene duplication leading to the multi-member IF family mentioned above. All intermediate filaments are formed from elongated non-globular monomers with a central rod domain, a globular carboxy terminal tail, and an amino terminal head domain. These monomers dimerise, form tetramers, and then, by a mechanism that is still unclear, form an IF. Regulation of IF disassembly/assembly is controlled by phosphorylation/dephosphorylation with the most dramatic example found in the nuclear lamins, which are hyperphosphorylated on their IF monomer amino head domain during mitosis by the mitosis specific CDC 2 kinase. Phosphorylation is also known to play a role in blocking neurofilament polymerisation.

Defining the functional roles played by IFs has proved difficult, although recent advances in molecular genetics and the use of transgenic mice have been productive. For example, antisense studies have unequivocally demonstrated that desmin filaments are involved in the myogenic differentiation that occurs when myoblasts fuse into the multinucleate myotubes that form skeletal muscle. Antisense oligonucleotides have also been used to show that GFAP is involved in the formation and maintenance of astrocyte processes. The role of vimentin has been analysed by using null mutation mice and, although these appear normal, it is possible that the role of the vimentin network is related to the function of other IF classes. For example, GFAP fails to form an organised network when injected into cells lacking vimentin. The reasons for this are presently unknown.

The role of neurofilaments (NFs) has been much more extensively studied because they are the major IF class in many types of mature neurons. Neurofilaments are assembled from three polypeptide subunits—NF-L, NF-M, and NF-H—and are abundant in axonal processes. Structural roles for them in establishing and/or maintaining neuronal asymmetry have been suggested for many years, and it is now known that synergistic interactions of the NF triplet is required for filament assembly, with a leading role played by NF-L and subordinate substoichiometric roles allocated to NF-M and NF-H. NF triplet gene expression is of considerable interest because it is elevated once neurons reach their target organs and axonal diameters increase by up to fivefold. Although suggestive, this relationship between axonal diameter and NF expression remained "not proven" until the identification of a mutant Japanese quail (quiver) in which neurofilaments were undetectable and axonal diameters were considerably reduced. Analysis of the quiver quail strongly suggests that the reason for this is a reduction, to less than 5% of normal levels, in the mRNA encoding the NF-L subunit. These data leave little doubt that neurofilaments are a key element in the radial growth of axons and that the neurological defect (quivering) is caused by a concomitant reduction in action potential velocity.

Intermediate filaments and disease

In this section only the keratins and NFs are discussed because, at present, there is little clearcut correlation between other IF classes and disease.

Keratin Epidermolysis bullosa simplex (EBS) is a hereditary skin disease which causes the formation of blisters that bulge upwards from the basal cell layer of epidermal cells found next to the dermis. These cells are unable to resist the normal mechanical trauma to which skin is subjected, break down, and, particularly in infants, can lead to severe secondary infection. It is now well established that EBS is caused by mutations in the keratin filaments (K5 and K14) of the basal cell layer which are polymerisation incompetent, fail to form filaments, and thus directly lead to the EBS pathogenesis of skin blistering and cell lysis. There is little doubt that, in this instance, mutated keratins are necessary and sufficient to cause this disease because transgenic mice expressing a truncated K14 keratin gene exhibit similar skin abnormalities.

Neurofilaments It has long been argued that NFs play a role in several neurodegenerative diseases such as the motor neuron diseases (MND), amyotrophic lateral sclerosis (ALS), infantile spinal muscular atrophy, and hereditary sensory motor neuropathy. These are diseases resulting from motor neuron failure, leading to skeletal muscle atrophy, paralysis, and death. A common feature of these diseases is the accumulation and aberrant assembly of NFs in motoneuron cell bodies and axons. This, of course, does not tell us whether NF accumulation has a causal role or is simply a byproduct of the onset of MND. Recent advances in molecular and cell biology have shown a partial way out of this apparent impasse. Transgenic mice in which NF-L expression was increased by four times the normal level were found to have increased amounts of NFs in motoneuron cell bodies and degenerating motor axons. More importantly, the skeletal muscle targets of these axons were atrophied, indicating that their motoneuron innervation was non-functional. Furthermore, the morphological effects of NF overproduction seen in these transgenic animals most clearly resembles those in MND. These data have clarified an important point, namely that a primary change in the NF cytoskeleton is both necessary and sufficient to produce many of the characteristics seen in the pathology of neurodegenerative MND. A major and thus far unanswered question is: how does the increase in NFs seen in MND cause neuronal cell death? Here too it seems likely that the powerful combination of molecular and cell biology will eventually provide the answer.

Actin and actin associated proteins

Actin is an abundant protein found in all eukaryotic cells at levels of up to 5% of total protein in non-muscle cells and 60% in skeletal muscle, where its role in muscle contraction is well known. Actin in non-muscle cells has many functions and is involved in controlling cell movement, cell division, cell shape, and the movement of cell surface receptors. Actin filaments are formed by the polymerisation of globular actin monomers—a process that, in vitro, requires physiological salt conditions (K^+ and Mg^{2+}) and ATP. Conversely, actin depolymerisation in vitro occurs in low salt (usually 2 mmol/l) Ca^{2+} and ATP. Each actin monomer interacts with another in a precisely defined and polarised way, and two strands of actin monomers interact with each other to form an actin filament. There are at least six different actins found in mammalian cells: class I comprising non-muscle β and γ plus smooth muscle γ, and class II containing cardiac α, skeletal α, and vascular α, all encoded by a related gene family. Actin is a highly conserved molecule differing by only a few amino acids between fungi and mammals. The ability of these separate actins to interact with each other in vivo was clearly demonstrated by transfecting the gene encoding cardiac α actin into fibroblasts (which do not express this actin isoform) and showing that cardiac actin can successfully co-polymerise into chimaeric filaments with the indigenous fibroblast β and γ actins. Myosin molecular motors are associated with actin filaments and provide the force for driving all cell movements; there are now at least eight myosins that can be divided into "conventional" bipolar myosins found in both muscle and non-muscle cells, together with myosin I, a unipolar myosin found only in non-muscle cells. Antisense and myosin gene knockout experiments using the slime mould *Dictyostelium* species have provided conclusive proof that myosin is essential for cell movement.

In addition to myosin, many other proteins interact with actin in non-muscle cells and control its polymerisation, organisation, and interactions. There are now at least 100 of these actin associated proteins (AAPs) which nucleate, sever, block polymerisation, depolymerise, stabilise, crosslink actin filaments to each other, and link them to transmembrane cell surface receptors by a molecular cascade of at least 10 AAPs. These molecules interact with actin using at least four defined actin binding domains—a number that seems certain to increase in the near future. In addition to its presence in the cytoplasm, it is of considerable interest that actin is also a significant component in the nucleus, where it is likely to form part of the nuclear skeleton. Nuclear actin is known to interact directly with chromosomes, and microinjection of anti-actin antibodies blocks transcription by RNA polymerase II, data that have been extended by the demonstration that the factor required for accurate initiation of RNA polymerase II transcription is actin. The amount of actin present in the nucleus is thought to be regulated by cofilin (an AAP), because actin does not possess the nuclear localisation signal necessary for transit from the cytoplasm into the nucleus. Although the function of nuclear actin is presently unclear, it may well involve the spatial organisation of chromosomes and the transcription of their genes. For example, dictyostelium ABP 50 (an actin binding protein) has been identified as elongation factor 1α.

These findings (and others) have led to the realisation that some components of the actin contractile system also have important roles in other cellular processes, such as positioning mRNA in the cytoplasm. For example, the anterior–posterior axis formation that occurs during development of *Drosophila* spp. defines the polarity of the whole organism, and is known to depend on the correct localisation of two mRNAs, bicoid and oskar to opposite ends of a single cell. Mislocalisation of either mRNA species causes pattern defects in the resulting embryos—a fault now believed to be caused by mutations in the gene encoding the AAP tropomyosin. The resulting reduction (or loss) of tropomyosin expression leads to actin filament destabilisation, loss of organisation, and this, in turn, is thought to perturb mRNA transport and cause incorrect development of *Drosophila* spp.

Actin, AAPs, and disease

The fundamental role played by actin and its AAPs in so many cell functions strongly suggests that any defect in the cytoskeletal network may well be manifested as disease. Although a direct cause and effect relationship is rare, however, it is not unknown, as illustrated by Duchenne muscular dystrophy and dystrophin.

Duchenne muscular dystrophy Proteins that link the cytoskeleton to transmembrane receptors act as transducers in a molecular cascade between the cytoskeleton and the extracellular matrix. An important example of such a linker protein, found in skeletal muscle, is dystrophin. It is now well established that the absence of dystrophin is both necessary and sufficient for the onset of Duchenne muscular dystrophy (DMD). This is an X chromosome linked, muscle wasting disease which affects about one in 3000 boys, and leads to death in the late teens/early twenties as a result of cardiac or respiratory failure caused by the necrosis of large skeletal muscle fibres. DMD is caused by mutations in the dystrophin gene (> 2·5 mega-bases with 79 axons) from which a 14 kilobase (kb) dystrophin mRNA is transcribed and a 427 kDa protein encoded. Dystrophin contains a binding site for actin filaments at its amino terminus and an additional binding site for a transmembrane complex of five glycoproteins at its carboxy terminus. The extracellular region of these glycoproteins binds α dystroglycan which, in turn, links to the extracellular matrix component laminin. This is how skeletal muscle fibres are securely attached to the extracellular matrix.

The critical question here is: how does the absence of a cytoskeletal associated protein (dystrophin) lead to muscle cell necrosis and DMD? The answer is almost certainly that the absence of dystrophin results in disruption of the linkage between the skeletal muscle cytoskeleton and the quintet of laminin binding transmembrane glycoproteins. This leads to contraction-induced muscle cell membrane damage followed by necrosis of skeletal muscle fibres. Dystrophin loss in DMD also leads to a downregulation in the expression of the transmembrane glycoproteins, further exacerbating muscle necrosis and the pathogenesis of DMD. Gene therapy for DMD has thus far proved difficult because of the large size of the mRNA that encodes dystrophin precluding successful expression vector packaging.

Bacterial and viral infection   Compelling evidence now exists that an intact actin cytoskeleton is necessary for cellular invasion by various bacteria and viruses. For example, the measles virus appears to express a nucleocapsid that nucleates actin and converts it from the globular to the filamentous form, which the virus can then use to bud from infected cells. Furthermore, it has been suggested that HIV uses the actin cytoskeleton of infected T lymphocytes to either form (or localise to) pseudopods (finger like extensions of the cell membrane) before infecting epithelial cells.

The role of actin in bacterial pathogenesis has been well defined recently, particularly in the case of *Listeria monocytogenes*. This organism moves in its host cell cytoplasm rapidly (at a rate of 6–60 μm/min compared with wound healing and metastasising tumour cells which move at 0·1–1 μm/h) by promoting continuous host cell actin assembly at one pole of the bacterium. When the bacterium reaches the host cell plasma membrane, actin polymerisation pushes out the membrane until it contacts the plasma membrane of a neighbouring cell. This finger like projection with a bacterium at the tip is then phagocytosed, the surrounding membrane lysed, and a new cycle of infection started. Thanks to its ability to "hijack" the host cell's actin, the parasite has therefore remained intracellular and successfully evaded the host's immune system. This can occur because *Listeria* expresses a phosphorylated protein called Act A which nucleates actin assembly into the filaments which then act as a "rocket" to propel the bacterium. The role of Act A has been investigated by expressing this gene, together with a carboxy terminal "anchor", in eukaryotic cells. This co-expressed "anchor" targets the Act A to mitochondrial membranes where actin filaments were recruited, thus showing directly how Act A functions. Furthermore, mutations of the Act A gene in *Listeria* itself totally prevented actin assembly and bacterial movement. These types of experiments show very clearly how molecular biology, in combination with cell biology, leads to enhanced understanding of disease.

Cancer One of the most important questions to be answered, after the diagnosis of cancer, is whether the disease is localised or has spread to sites distant from its origin. The enhanced movement of malignant tumour cells that invade and destroy normal tissue architecture is a major contributor to the progression of this disease, with many cells found in secondary tumours at sites distant from their origin. To metastasise, a tumour cell must initially dissociate from the primary tumour (with a concomitant loss of cell–cell and cell–matrix adhesions), migrate through connective tissue and capillary walls into the circulatory system, extravasate, again migrate through tissue, and form a secondary tumour. These oncogenically activated cell movements and alterations in cell shape require spatially and temporally regulated protease secretion, alteration of integrin expression, modification of integrin–actin linkage proteins, and the coordinated activation of the actin cytoskeleton, resulting in the extension of invadopodia and cell locomotion.

Thus, the actin cytoskeleton plays an important role in metastasis with oncogenic transformation affecting both actin and its associated proteins. Many normal tissue associated cells contain bundles of actin in the form of stress fibres that cross the cytoplasm. These fibres are associated with anchorage dependence, reduced mobility, and contact inhibition. In motile cells and many transformed mesenchymal cells, however, actin stress fibre bundles are absent and actin filaments are distributed diffusely through the cytoplasm or concentrated in the cortex where they form a meshwork of randomly crosslinked filaments. Downregulation of the α actin isoform, mutated actin isoform expression, and imbalances in actin isoform expression are all associated with, and very probably important in, oncogenically induced cell shape change and activated cell movement. Direct evidence for the role of the actin network in metastasis comes from experiments showing that the invasive capability of melanoma cells is decreased after transfection of the gene encoding a mutated form of β actin, possibly as a result of this actin isoform forming more actin stress fibre bundles in these cells.

Several AAPs are now known to be affected by oncogenesis but, for the purposes of this chapter, those discussed are those where reversal of the transformed phenotype occurs after gene transfection with an AAP or where the gene involved in tumour formation encodes an AAP. Examples include the actin crosslinking and plasma membrane linking protein α actinin, which is expressed at very low levels in SV40 transformed fibroblasts. These cells have a more rounded morphology than their normal counterparts, grow in soft agar, and form tumours in vivo. Transfection of the gene encoding α actinin results in suppression of the transformed phenotype, a finding that may reflect either the stabilising effect this protein has on actin filament bundles or its role as a link between the actin cytoskeleton and transmembrane receptors.

Multiple isoforms of tropomyosin, an AAP that stabilises actin filaments and regulates contraction, are found in non-muscle cells, but only the higher molecular weight isoforms are downregulated in transformed cells. Transfection of a single isoform only partly reverses tumorigenicity, suggesting that restoration of a "functional unit" of these proteins is required (see also below). It is of considerable interest that the 3' untranslated region of the α tropomyosin mRNA is also effective in suppressing tumorigenicity, but how this occurs is presently unclear. Evidence directly linking an AAP with the onset of cancer is the finding that the tumour suppressor gene inactivated in neurofibromatosis encodes schwannomin—an actin–plasma membrane linkage molecule. It has been suggested that the loss of schwannomin may lead to the altered cell–cell, cell–matrix interactions and tumour cell migration that parallel the development of central nervous system tumours.

At present, there are no clearcut rules regarding the effect that oncogenesis has on AAPs, because the expression level of one oncogenically modified AAP can vary from one tumour to another, even within the same type of tumour. This apparent inconsistency may well be because we have not yet identified all the molecules involved or it may reflect a highly complex control mechanism in which any perturbation of actin/AAP expression levels unbalances the delicate equilibrium that controls normal cell behaviour. Oncogenically induced, actin based, cell movements can be activated by alterations to either a single AAP or a synergistic functional unit of these molecules. This is very likely to reflect the apparent functional overlap between these proteins that is necessary for them to provide an essential molecular backup system for the cell. However, in some instances the transfection and re-expression of a single protein (for example, α actinin or vinculin) in a tumour cell can rescue the transformed phenotype, suggesting the presence of at least two molecular pathways, one of which, possibly involving the actin–transmembrane linkage, is not underwritten by other AAPs.

As the movement of metastatic cells resembles that of many normal cells during embryogenesis, tissue repair, and the invasion of sites of infection by cells of the immune system, it is unlikely to be the loss of a functional actin cytoskeleton, but its continual activity, that is important in oncogenesis. These activated movements are normally repressed by AAPs which are, therefore, acting as a form of tumour suppressor, and their altered expression removes a series of molecular brakes, resulting in the oncogenically activated cell migration that characterises metastatic cells.

Selected reading

Stossel TP. On the crawling of animal cells. *Science* 1993;**260**:1086–93.

Microtubules and microtubule associated proteins

Crowther RA. Steps towards a mouse model of Alzheimer's disease. *Bioessays* 1995; **17**:593–5.

Hoyt MA. Cellular roles of kinesin and related proteins. *Curr Opin Cell Biol* 1994; **6**:63–8.

Mandelkow E, Mandelkow E-M. Microtubules and microtubule associated proteins. *Curr Opin Cell Biol* 1995;**7**:72–81.

Intermediate filaments

Heins S, Aobi U. Making heads and tails of intermediate filament assembly, dynamics and networks. *Curr Opin Cell Biol* 1994;**6**:25–33.

Klymkowsky M. Intermediate filaments: new proteins, some answers, more questions. *Curr Opin Cell Biol* 1995;**7**:46–54.

Lee MK, Cleveland DW. Neurofilament dysfunction: involvement in axonal growth and neuronal disease. *Curr Opin Cell Biol* 1994;**6**:34–40.

McLean WHI, Lane EB. Intermediate filaments and disease. *Curr Opin Cell Biol* 1995;**7**:118–25.

Actin and actin associated proteins

Button E, Shapland C, Lawson D. Actin: its associated proteins and metastasis. *Cell Motil Cytoskel* 1995;**50**:247–51.

Campbell KP. Three muscular dystrophies: loss of cytoskeleton—extracellular matrix linkage. *Cell* 1995;**80**:675–9.

Cossart P. Actin based bacterial mobility. *Curr Opin Cell Biol* 1995;**7**:94–101.

Janmey PA, Chaponnier C. Medical aspects of the actin cytoskeleton. *Curr Opin Cell Biol* 1995;**7**:111–17.

Pollard TD. Actin and actin binding proteins. In Kreis T, Vale R, eds. *Guidebook to the cytoskeletal and motor proteins.* Oxford: Oxford University Press, 1993;3–11.

Spudich JA. How molecular motors work. *Nature* 1994;**372**:515–18.

12: The cell nucleus

R A Laskey

The cell nucleus is the information centre of the cell. It is responsible for copying highly selected regions of the genome into ribonucleic acid (RNA) and for supplying precisely regulated amounts of specific RNA molecules to the cytoplasm, where they are translated into proteins. In addition it must duplicate its entire structure every time the cell divides.

This chapter considers how the nucleus is organised to perform the immensely complex task of selective information retrieval. Recently, several different experimental approaches have elucidated how deoxyribonucleic acid (DNA) is organised in the nucleus and how specific regions of DNA are selected for expression at particular times and in particular cells of the body. One of the key problems in selective information retrieval in the nucleus becomes obvious if we consider how densely DNA is packed into the nucleus. Each chromosome consists of an individual double helix of DNA about 40 mm long by 2 nm wide packed into a nucleus of about 6 µm in diameter. The implications of these dimensions can be seen more clearly by a simple scale model enlarged one million times. On this scale the DNA in each chromosome would resemble thin string, 2 mm in diameter but 40 km long, and the total DNA in one nucleus would reach from London to Naples. Yet on the same scale this 2000 km of DNA must all be packaged into a nucleus of only 6 m diameter in such a way that specific regions must remain fully accessible for highly regulated expression.

The answer to how such a densely packed structure can function in selective information retrieval lies in a precise three dimensional nuclear architecture. The contents of the nucleus are not just a randomly mobile solution, but are highly organised into a hierarchy of ordered structures.

Chromatin and chromosomes

The fundamental unit of DNA packaging in higher organisms is the nucleosome, in which 146 base pairs of DNA are wrapped twice around an octamer of histones, consisting of two each of histones H2A, H2B, H3, and H4. These small basic proteins neutralise the acidic charges of DNA and they also shorten it by providing spools on which the DNA is coiled.

A fifth histone, called H1, occupies the site at which DNA enters and leaves the nucleosome. When H1 is present nucleosomes can coil into a solenoid with six nucleosomes in each turn. This has two important consequences. First, coiling compacts DNA so that it can be divided between progeny cells at mitosis and, second, it allows selected regions of DNA to be folded away so that only a specific subset of the genetic information is available for expression in a particular type of cell. Active genes are packaged into a different, and more accessible, chromatin conformation from that of inactive genes in the same cell. Thus nucleosomes allow a structural level of gene regulation, by allowing selective access to regions of DNA. In addition, the absence of a nucleosome or a small group of nucleosomes from the region of DNA just upstream of a gene can be important in gene regulation. These regions are hypersensitive to nucleases such as DNase 1. They frequently correspond to regions of DNA that are essential for correct gene regulation (see below). Proteins are emerging that displace nucleosomes during gene expression.

Not only is the DNA in each chromosome packaged into nucleosomes, it is also organised into a series of loops which extend radially from an axial scaffold. Loops vary in size but on average they contain 20 000–100 000 base pairs each. Specific proteins bind to and grip DNA sequences which define the bases of these loops. This level of organisation persists when chromosomes are condensed for cell division or when they are decondensed for transcription and replication during interphase. Furthermore, the organisation does not appear to change when cells differentiate. Instead it appears to define a basal level of organisation within the nucleus (fig 12.1).

Each chromosome consists of a linear chain of genes, but it also contains other structural features, including centromeres and telomeres. Centromeres attach chromosomes to the spindle during mitosis and meiosis, whereas telomeres form the termini of chromosomes. Both centromeres and telomeres have been isolated

141

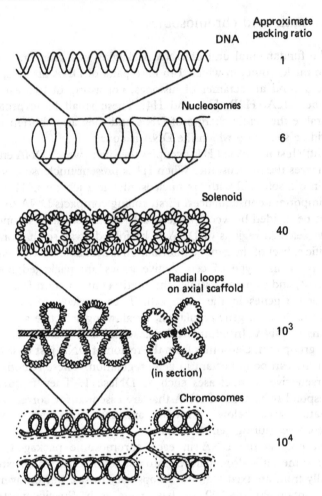

DNA — Approximate packing ratio: 1

Nucleosomes — 6

Solenoid — 40

Radial loops on axial scaffold — 10^3

(in section)

Chromosomes — 10^4

FIG 12.1—*Summary of structural models for the packing of DNA into metaphase chromosomes. (Reprinted with permission from Laskey, 1986.)*

and characterised by selecting for these functions on artificial chromosomes in yeast. Centromeres allow plasmids to segregate stably to progeny cells during mitosis and meiosis, whereas telomeres allow linear DNA molecules to complete their replication correctly. DNA polymerases are unable to replicate right to the ends of a linear chromosome, so without a specialised telomere, chromosomes would become slightly shorter each cell generation. The telomeric DNA sequence is a simple repeat which can fold back to form a hairpin loop or even a four stranded structure involving hydrogen bonding between four deoxyguanosine residues to form "G quartets". These simple repeats are not synthesised by copying the parental DNA strands, but by an enzyme called telomerase which contains a short RNA template. Telomerase copies this RNA template into DNA, extending the ends of the chromosomes.

The ability to construct artificial chromosomes in yeast has allowed components essential for chromosome function to be identified. Apart from centromeres, telomeres, and genes, yeast chromosomes also require specific DNA sequences to serve as initiation sites for DNA synthesis (origins of replication). Obviously we would like to know more about equivalent sequences from mammalian cells. Human telomeres have been isolated and they are strikingly similar to those of yeast, but human centromeres and replication origins are proving to be much more complex.

Imports and exports through the nuclear envelope

The nucleus is surrounded by two complete layers of membrane, but these do not exclude small molecules because the membranes are perforated at frequent intervals by nuclear pore complexes. These are grommet like structures which allow the free passage of small molecules up to the size of small proteins, although they regulate selectively the transit of larger molecules in both directions. Thus, above about 50 kilodaltons in molecular weight, only "nuclear" proteins can enter the nucleus, and other proteins are excluded.

The information which specifies protein entry into the nucleus resides in very short stretches of amino acids, examples of which are given in table 12.1. Clusters of positively charged amino acids are found, but neutral and even acidic amino acids can be important, as in the case of the c-myc signal. When a short synthetic peptide of such a sequence is crosslinked to cytoplasmic proteins it causes them to accumulate in the nucleus. The nuclear pore has been identified as the route of protein entry by coating colloidal gold particles with a nuclear protein, and using electron microscopy to follow their fate after injection into cytoplasm. The gold particles can be seen aligned through the centre of the nuclear pore in figure 12.2.

Nuclear import signals (also called nuclear localisation signals or NLSs) are recognised in the cytoplasm by a soluble receptor

TABLE 12.1—*Signals for nuclear import or export*

Nuclear import signals

SV40 T antigen	Pro	Lys(+)	Lys(+)	Lys(+)	Arg(+)	Lys(+)	Val		
Nucleoplasmin	Lys(+)	Arg(+)–10 amino acids–Lys(+)			Lys(+)	Lys(+)	Lys		
c-myc	Pro	Ala	Ala	Lys(+)	Arg(+)	Val	Lys(+)	Leu	Asp

Nuclear export signals

HIV-1 Rev	**Leu**	Pro	Pro	**Leu**	Glu	Arg	**Leu**	Thr	**Leu**
Protein kinase inhibitor	**Leu**	Ala	**Leu**	Lys	**Leu**	Ala	Gly	**Leu**	Asp	Ile

Import signals are characterised by essential positively charged amino acids (+), whereas the nuclear export signals mapped to date are characterised by clusters of hydropholic leucine residues (**bold**).

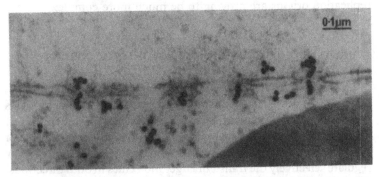

FIG 12.2—*Electron micrograph of a section through the nuclear envelope of a frog oocyte which has been injected with colloidal gold particles coated with the nuclear protein "nucleoplasmin". Gold particles can be seen aligned through nuclear pore complexes. (Courtesy of AD Mills.)*

called importin (fig 12.3). The α subunit of importin binds the import signal in the cytosol. The β subunit then binds to the nuclear pore, causing its nuclear protein passenger to dock at the pore. A small GTP binding protein called Ran is then required for the act of translocation through the nuclear pore. The α and β importins stay together with their passenger protein until they reach the inside of the nuclear pore. There they dissociate, α entering the nucleoplasm and β remaining at the inner face of the pore complex.

Recent evidence implicates the α subunit of importin in the export of RNA from the nucleus, although several RNA binding proteins also have roles in export. The nuclear pore is also known

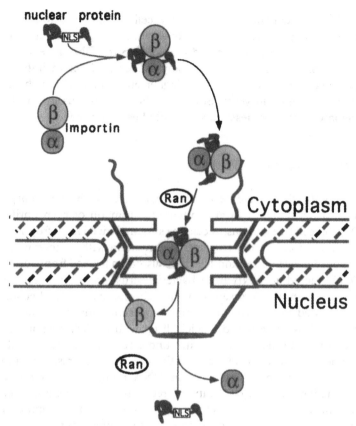

FIG 12.3—*Schematic summary of the nuclear protein import pathway. Importin α binds the nuclear localisation signals of nuclear proteins in the cytosol. Importin β docks at the nuclear pore complex. Ran allows transit to the inside of the pore where the α and β subunits of importin dissociate; α enters the nucleoplasm whilst β returns from the pore to the cytoplasm. (Reproduced from Laskey et al, (1996), with permission.)*

to be the route of export. This process is also highly selective, resulting in the export of mature transcripts, but the retention in the nucleus of immature transcripts.

Functions of the cell nucleus

Major functions of the cell nucleus include the synthesis, processing, and export of RNA to the cytoplasm. These topics are covered in chapter 13 and are not duplicated here.

The second main function of the cell nucleus is chromosome replication. Eukaryotic chromosomes initiate replication at many sites, yet these are coordinated so that all of the DNA is replicated exactly once and only once in any cell cycle. Sites of initiation have been difficult to define in higher eukaryotes, although they are well defined in yeast. Several of the key proteins involved in regulation and synthesis have recently been identified.

Rebuilding the cell nucleus

Recently there have been substantial advances in our ability to reconstruct "nuclei" from condensed chromatin or even purified DNA in cell free systems. A cell free system derived from frog's eggs assembles nuclei from frog sperm chromatin with its membranes removed in vitro. Pseudonuclei have also been formed from purified DNA, including DNA from bacterial viruses and plasmids. Not only is such DNA assembled into chromatin, it also becomes surrounded by nuclear membranes containing functional nuclear pores. Thus these pseudonuclei fill up with nuclear proteins and even replicate their DNA in vitro under strict cell cycle control.

There is growing evidence that the structural organisation of the nucleus plays important roles in DNA replication and mRNA production. The opportunity to reconstruct functioning nuclei and functioning chromosomes opens up new opportunities to understand how these complex structures perform their extraordinary functions.

Further reading

Alberts B, Bray D, Lewis J, Raff M, Roberts K, Watson JD. *Molecular biology of the cell*, 3rd edn. New York: Garland, 1994: chaps 8, 9, and 11.

Coverley D, Laskey RA. Regulation of eukaryotic DNA replication. *Annu Rev Biochem* 1994;**63**:745–76.

Dingwall C, Laskey RA. Nuclear targeting sequences—a consensus? *Trends Biochem Sci* 1991;**16**:478–81.

Görlich D, Mattaj IW. Nucleocytoplasmic transport. *Science* 1996;**271**:1513–8.

Laskey RA. Prospects for reassembling the cell nucleus. *J Cell Sci* 1986;4(suppl): 1–9.

Laskey RA, Görlich D, Madinè MA, Makkesh JPS, Romanowski P. Regulatory roles of the nuclear envelope. *Exp Cell Res* 1996;**229**:240–11.

13: Gene regulation and transcription factors

David S Latchman

It should be clear to anyone who has ever dissected a human body that the expression of human genes must be a highly regulated process. The vast range of different tissues and organs differ dramatically from each other and they all synthesise different proteins—haemoglobin in red blood cells, myosin in muscle, albumin in the liver, and so on. Moreover, with few exceptions all these different cell types contain the same sequence of DNA, which encodes all these different cell proteins, and this DNA is also identical to the DNA in the single celled zygote, from which all these different cells arise during embryonic development. Clearly, therefore, some process of gene regulation must operate to decide which genes within the DNA will be active in producing proteins in each cell type.

Levels of gene regulation

A number of stages exist between the DNA itself and the production of a particular protein (fig 13.1).[1][2] Thus the DNA must first be transcribed into a primary RNA transcript, which is subsequently modified at both ends by addition of a 5′ cap and a 3′ tail of adenosine residues. Moreover, within this primary transcript, the RNA sequences that actually encode the protein are not present as one continuous block. Rather they are broken up into segments (exons) which are separated by intervening sequences (introns) that do not contain any protein coding information. As these introns interrupt the protein coding region and would prevent the production of an intact protein, they must be removed by the

147

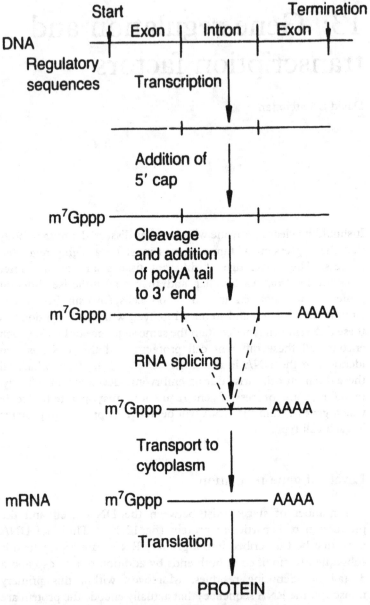

FIG 13.1—*Stages in gene expression which could be regulated.*

process of RNA splicing[3] before the mature messenger RNA can be transported from the nucleus to the cytoplasm and translated into protein.

Clearly each of these stages is a potential point at which gene expression could be regulated, and there is evidence that several of them are actually used. Thus, for example, the production of many new proteins in the egg immediately after fertilisation and the start of embryonic development depend on the translation into protein of fully spliced, messenger RNAs that pre-existed in the cytoplasm of the unfertilised egg but the translation of which was blocked before fertilisation. This form of gene regulation is known as translational control. Similarly, by splicing the protein coding regions (exons) of a single primary transcript in different combinations, two or more different mRNAs encoding different proteins in different tissues can be produced. This process of alternative splicing[4] is well illustrated in the single gene that encodes both the calcium modulating hormone, calcitonin, and the potent vasodilator, calcitonin gene related peptide (CGRP) (fig 13.2). Thus this gene is transcribed into a primary RNA transcript both in the thyroid gland and in the brain, but a different combination

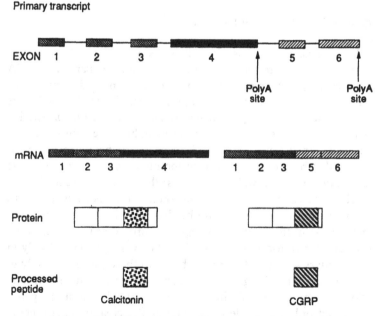

FIG 13.2—*Alternative splicing of primary transcript of calcitonin/calcitonin gene related peptide gene in brain and thyroid cells. Splicing followed by proteolytic cleavage of precursor protein in each tissue yields calcitonin in thyroid and calcitonin gene related peptide in brain.*

of exons is spliced together in each cell type to produce calcitonin mRNA in the thyroid and CGRP mRNA in the brain.

Although there are thus several cases of gene regulation after the first stage of transcription, a wide variety of evidence indicates that, in most cases, gene regulation is achieved at the initial stage of transcription, by deciding which genes should be transcribed into the primary RNA transcript.[2] In these cases, once transcription has occurred, all the other stages in gene expression shown in figure 13.1 follow, and the corresponding protein is produced. Thus the myosin gene is transcribed only in muscle cells, resulting in myosin being produced only in this cell type; the immunoglobulin gene is transcribed only in B lymphocytes, which produce immunoglobulin; and so on. Indeed, even in the case of calcitonin and CGRP, alternative splicing is acting as a supplement to transcriptional control, because calcitonin/CGRP gene is transcribed only in the thyroid gland and the brain and not in other tissues. Thus the regulation of gene transcription has a critical role in the regulation of gene expression.

Regulation of transcription

For a gene to be transcribed, it is necessary for specific protein factors known as transcription factors[5] to bind to particular DNA binding sites in the regulatory regions of the gene and to induce its transcription by the enzyme RNA polymerase. Some of these factors are present in all cell types whereas others are active only in specific cells or after exposure to a particular stimulus. The combination of particular binding sites in a particular gene determines the transcription factors that bind to it, and in turn the presence or absence of these factors determines in which cell type(s) the gene is transcribed.

Thus, for example, the genes encoding proteins such as growth hormone and prolactin contain binding sites for the transcription factor Pit-1 in their regulatory region or promoter upstream of the start site for transcription. The Pit-1 factor is synthesised only in the pituitary gland and therefore binds to these promoters only in pituitary cells, resulting in the transcription of the growth hormone and prolactin genes only in the pituitary gland.[6] Similarly, genes expressed only in muscle cells, such as the creatine kinase gene, contain binding sites for the MyoD transcription factor, which is present only in muscle cells. This case is even more dramatic, however, because the artificial expression of MyoD in non-muscle

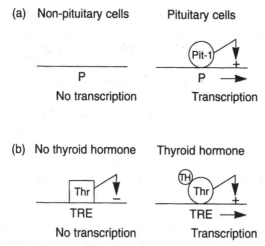

(a) Non-pituitary cells   Pituitary cells

P
No transcription

P →
Transcription

(b) No thyroid hormone   Thyroid hormone

TRE
No transcription

TRE →
Transcription

FIG 13.3—*Regulation of transcription factors at the level of synthesis (a) or activity (b). (a) The regulation of the Pit-1 transcription factor at the level of its synthesis. The factor is absent in non-pituitary cells and is synthesised only in pituitary cells. It therefore binds to its binding site (P) in the DNA of its target gene only in pituitary cells and so these genes are only transcribed in such cells. (b) The regulation of transcription factor activity in the case of the thyroid hormone receptor (Thr). In the absence of thyroid, the receptor binds to its binding site (TRE) in the DNA of its target genes but cannot stimulate transcription. Indeed it actually has an inhibitory effect on transcription. After the binding of thyroid hormone (TH) the receptor undergoes a conformational change which allows it to bind thyroid hormone and stimulate transcription of the target genes which are therefore stimulated by the presence of thyroid hormone.*

cells such as fibroblasts is sufficient to convert them into muscle cells, indicating that MyoD activates transcription of all the genes whose protein products are necessary to produce a differentiated muscle cell.[7]

In these examples, therefore, the activity of the transcription factor is regulated at the level of synthesis by ensuring that it is made only in the cell types where it is needed to activate gene expression. This is illustrated for the case of Pit-1 in figure 13.3a. In addition, however, transcription factors can also be regulated at the level of activity, being synthesised in a variety of cell types but undergoing conversion to a form that can stimulate gene transcription only in a specific cell type or after exposure to a specific signal.

An example of a factor regulated in this manner is NFκB which binds to a DNA sequence in the regulatory region of the immunoglobulin κ light chain gene. Interestingly the NFκB protein is present in all cell types. In most cells, however, it is present in an inactive form in which it is complexed with an inhibitory protein, resulting in it being restricted to the cell cytoplasm. In mature B cells, however, NFκB is released from the inhibitory protein and moves to the nucleus, where it can bind to its DNA target sequence and activate the transcription of the immunoglobulin κ light chain gene.[8] Interestingly, this activation of NFκB also occurs when resting T lymphocytes are activated by antigenic stimulation, and is the main reason for the improved growth of the human immunodeficiency virus (HIV) in activated compared with resting T cells, because NFκB can bind to two sites within the HIV promoter and activate viral transcription.

In this case therefore gene expression is regulated by controlling the ability of NFκB to bind to the appropriate DNA sequence. In addition, however, regulation of transcription factor activity can also occur after the factor has bound to the DNA. This is seen in the case of the thyroid hormone receptor which mediates the cellular response to thyroid hormone. Before exposure to the hormone, the receptor is already bound to its corresponding DNA sequence in its target genes. However, its structure is such that it cannot exert a stimulatory effect on transcription and indeed it actually inhibits the rate of transcription. After exposure to thyroid hormone, the receptor undergoes a conformational change which now allows it to activate transcription. Hence in this case transcription factor activity is regulated at the level of its ability to stimulate transcription rather than at the level of DNA binding ability[9] (fig 13.3b).

Hence the effect of transcription factors on gene expression can be controlled both by the regulation of their synthesis (as illustrated for Pit-1 in fig 13.3a) and by the regulation of their activity (as illustrated for the thyroid hormone receptor in fig 13.3b). The combination of these two processes allows transcription factors to regulate the expression of numerous different genes in different cell types.

Malregulation of gene expression in disease

Mutations in binding sites for regulatory proteins

In view of the complex nature of gene regulation, it is not surprising that it can go wrong and that a number of human diseases have now been shown to be the result of defects in gene regulation. In some cases of this type, the disease arises from a mutation in a particular DNA sequence which is the target for a regulatory protein. Thus, in haemophilia B a mutation occurs in the promoter of the factor IX gene, resulting in the failure of a specific transcription factor to bind to this sequence. As the binding of this factor to the promoter of the factor IX gene is essential for its transcription, this results in a failure of factor IX gene transcription leading to disease.[10] Similarly mutation in the porphobilinogen deaminase gene results in a failure to produce one of two alternately spliced mRNAs derived from this gene, resulting in one form of acute intermittent porphyria.[11]

Transcription factor mutations

As well as mutations in binding sites for regulatory proteins, specific diseases can also result from the inactivation of a transcription factor by an inherited mutation. It might be thought that the vital role of these factors, together with the fact that a single factor can affect the expression of many target genes, would result in such inactivation of a specific factor being incompatible with survival. Although this is probably true in many cases, mutations in some transcription factors are compatible with survival and result in specific diseases. Thus, for example, in the case of the PAX family of transcription factors,[12] mutations in specific members of the family have been shown to result in Waardenburg's syndrome (PAX3), optic nerve colobomas and renal abnormalities (PAX2), and eye defects such as aniridia and Peters' anomaly (PAX6) (see chapter 15).

Similarly mutations in the Pit-1 factor described above have been identified in patients with combined pituitary hormone deficiency in which an absence of growth hormone, prolactin, and thyroid stimulating hormone results in mental handicap and growth deficiency.[13] Interestingly, the mutant Pit-1 factor in these patients is still capable of binding to its DNA binding site in the target genes but, unlike the normal factor, does not activate transcription after such binding. This means that the mutant Pit-1 protein not only fails to stimulate gene expression but can also inhibit gene activation by preventing the normal protein from binding to the DNA (fig 13.4a). These findings thus explain why such mutations are dominant producing the disease even when a single copy of the wild type gene encoding functional Pit-1 is also present.

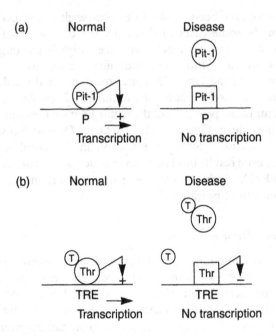

FIG 13.4—*Effects of transcription factor mutations in disease. (a) The production of a mutant Pit-1 protein (□) which can bind to its appropriate DNA binding site (P) but cannot stimulate transcription. It therefore acts in a dominant manner by preventing the binding of functional Pit-1 (○) resulting in a failure of transcription of Pit-1 dependent genes. In (b), a mutant thyroid hormone receptor (□) is able to bind to its specific binding site (TRE) but cannot bind thyroid hormone. It therefore fails to undergo the appropriate conformational change (see fig 13.3b) which occurs upon exposure to thyroid hormone. In addition, it also blocks the binding of functional thyroid hormone receptor (○) which has bound thyroid hormone and could stimulate transcription. This results in a failure of transcription of thyroid hormone dependent genes.*

Such dominant negative mutations which inhibit the function of the normal gene product are also seen in factors that are regulated at the level of activity. Thus mutations in the thyroid hormone receptor can result in the syndrome of generalised thyroid hormone resistance, which results from a failure to respond to thyroid hormone. The mutant receptors in this case are capable of binding to DNA and inhibiting transcription. Unlike the normal receptor, however, they cannot bind to thyroid hormone and therefore do not undergo the conformational change that allows the receptor to be activated after binding of thyroid hormone. They do not therefore allow activation by the activated normal receptor after exposure to thyroid hormone, and hence also act as dominant negative factors[14] (fig 13.4b).

Proto-oncogenes and anti-oncogenes

As well as cases such as these, where a disease is caused by the failure to express a particular gene, malregulation of gene expression can also result in disease if it causes genes to be expressed at the wrong time or in the wrong place. This form of malregulated gene expression is central to the development of certain cancers. Thus, it is now clear that many human cancers are caused by the mutation or overexpression of certain specific cellular genes known as proto-oncogenes, which results in their conversion into cellular oncogenes capable of causing cancer[15] (see chapter 14).

Although proto-oncogenes encode many different types of cellular proteins capable of stimulating cellular growth, such as growth factors or their receptors, several, such as *erbA*, *fos*, *jun*, *myb*, and *myc*, encode cellular transcription factors that are involved in regulating the expression of specific genes. After the conversion of these proto-oncogenes into oncogenes by mutation or overexpression, corresponding alterations occur in the expression of the genes that they regulate, resulting in cancer.

Thus, for example, in humans alterations in the expression or activation of a specific oncogene encoding a transcription factor have now been shown to be the cause of many different types of leukaemias, which were initially shown to be characterised by a particular chromosomal translocation.[16] Thus, for example, in Burkitt's lymphoma a portion of chromosome 8 containing the *myc* oncogene is relocated to chromosome 14 bringing the *myc* gene adjacent to the immunoglobulin heavy chain gene and resulting in greatly enhanced expression of the *myc* gene. As this gene is normally expressed only transiently in growing cells where it functions to stimulate proliferation, its greatly enhanced expression results in rapid growth of the cells leading to leukaemia. As well as such increases in expression, chromosomal translocations can also result in the production of a fusion gene composed of portions of two different genes previously present on separate chromosomes. In some cases such rearrangements result in leukaemia, presumably because the fusion protein has oncogenic properties distinct from that of either protein alone. This is seen, for example, in acute promyelocytic leukaemia where the gene encoding retinoic acid receptor α (which is related to the thyroid hormone receptor) is fused to another gene encoding the transcription factor for promyelocytic leukaemia.

Chromosomal translocations resulting in the enhanced expression or mutation of potentially oncogenic transcription factors thus play a key role in many human leukaemias and have also been described in solid tumours. In addition, however, human cancers can also result from alterations in another class of genes which encode proteins that restrain rather than stimulate cellular growth. These genes are known as anti-oncogenes.[17] Obviously, in such cases cancer will result from mutations that inactivate the anti-oncogene resulting in a failure to restrain cellular growth. As with the oncogenes, several anti-oncogenes have now have been shown to encode transcription factors. These include the retinoblastoma and Wilms' tumour genes, which were discovered on the basis of their germline mutation in patients with these tumours as well as the *p53* gene which is mutated in a wide range of human tumours.

Together, therefore, inactivating mutations in anti-oncogenic transcription factors and enhanced expression or activating mutations in oncogenic transcription factors play a key role in a wide variety of human cancers. When taken together with the role of transcription factor mutations in other diseases involving aberrant development or abnormal response to hormones, it is clear that alterations in the genes encoding transcription factors play a critical role in a wide variety of human diseases.[18]

Conclusion

This review gives a brief overview of the major aspects of gene regulation mechanisms (for further details see Latchman[25 18]). But much remains to be understood. For example, it is still unclear how the expression of specific genes is regulated both spatially and temporally during development, so that each cell type arises in the correct place and at the correct time. It is already clear, however, that the correct regulation of gene expression is central to health and correct development and that its malregulation is involved in a number of diseases.

1 Nevins JR. The pathway of eukaryotic mRNA transcription. *Annu Rev Biochem* 1983;**52**:441–6.

2 Latchman DS. *Gene regulation: A eukaryotic perspective*, 2nd edn. London: Chapman & Hall, 1995.

3 Sharp PA. Splicing of messenger RNA precursors. *Science* 1987;**235**:766–71.

4 McKeown M. Alternative mRNA splicing. *Annu Rev Cell Biol* 1995;**8**:133–55.

5 Latchman DS. *Eukaryotic transcription factors*, 2nd edn. London: Academic Press, 1995.

6 Anderson B, Rosenfeld MG. Pit-1 determines cell types during development of the anterior pituitary gland. *J Biol Chem* 1994;**269**:29335–8.

7 Edmondson DG, Olson EN. Helix–loop–helix proteins as regulators of muscle-specific transcription. *J Biol Chem* 1993;**268**:755–8.

8 Thanos D, Maniatis T. NFκB: a lesson in family values. *Cell* 1995;**80**:529–32.

9 Baniahmad A, Kohne AC, Renkawitz R. A transferable silencing domain is present in the thyroid hormone receptor, in the v-erbA oncogene product and the retinoic acid receptors. *EMBO J* 1992;**11**:1015–23.

10 Crossley M, Brownlee GG. Disruption of a C/EBP binding site in the factor IX promoter is associated with haemophilia B. *Nature* 1990;**345**:444–6.

11 Grandchamp B, Picat C, Mignotte V, Wilson JHP, Te Velde K, Sandkuyl L, *et al.* Tissue-specific splicing mutation in acute intermittent porphyria. *Proc Natl Acad Sci USA* 1989;**86**:661–4.

12 Hill RE, Hanson MJ. Molecular genetics of the Pax gene family. *Curr Opin Cell Biol* 1992;**4**:967–72.

13 Radovick S, Nations M, Du Y, Berg LA, Weintraub BD, Wondisford FE. A mutation in the POU-homeodomain of Pit-1 responsible for combined pituitary hormone deficiency. *Science* 1992;**257**:1115–18.

14 Baniahmad A, Tsai SY, O'Malley BW, Tsai MJ. Kindred S thyroid hormone receptor is an active and constitutive silencer and a repressor for thyroid and retinoic acid responses. *Proc Natl Acad Sci USA* 1992;**89**:10633–7.

15 Bourne HR, Varmus HE. Oncogenes and cell proliferation. *Curr Opin Genet Devel* 1992;**2**:1–57.

16 Rabbits TH. Chromosomal translocations in human cancer. *Nature* 1994;**372**:143–9.

17 Weinberg R. Tumour suppressor genes. *Neuron* 1993;**11**:191–6.

18 Latchman DS. Transcription factor mutations and disease. *N Engl J Med* 1996;**234**:28–33.

14: Genes and cancer

Richard G Vile, Myra O McClure, Jonathan N Weber

It is clear that, during development from a single cell zygote to a multicellular organism, a critical balance must be maintained between cell numbers (the ability of individual cells to proliferate) and cell specialisation (their concomitant ability to differentiate). Without sufficient levels of proliferation there will be too few cells to carry out the functions specific to their lineage; without appropriate degrees of differentiation within the lineage the functional specialisation of the tissue will be impossible to maintain. The ability of a cell to proliferate is intricately coupled with, and generally inversely related to, its ability to differentiate, as each phenotype is ultimately determined by activation of separate programmes of gene expression. It is from disruption of this fine balance of proliferative and differentiative genetic programmes that tumours inevitably arise.

A cancer may be viewed as a population of cells that has progressed a certain distance in its maturation pathway, but in which the processes of proliferation and differentiation have become uncoupled, and that, critically, is no longer able to complete its programme of differentiation. This population of cells is usually derived from divisions of a single cell that has acquired an accumulation of damaging changes in its genes. Cancer cells are said to express the fully transformed phenotype when they have acquired the ability to grow continuously, free from the normal inhibition usually imposed by their nearest neighbours (contact inhibition), and when they have become malignant. A malignant tumour consists of fully transformed cells that can invade adjacent tissues and spread (metastasise) to other sites in the body to form secondary tumour growths. Most cells can be diverted from their normal differentiation programmes within most of the compartments of the body and at many different stages between stem cell and the fully differentiated state.

Cancer as a disease of genes

For many years it has been realised that damage to the DNA of a cell (mutation) is associated with the changes that lead to cancer (carcinogenesis). At the turn of the century, Boveri identified the chromosomes as the storage place of the cell's genetic material and proposed chromosome imbalance as a major contributory factor to cancer development.[1] It has been shown repeatedly that most carcinogens are also active mutagens and the ability to cause damage to DNA correlates in most cases very well with the ability to induce cancer. Up to 70% of human cancers result from the action of chemical carcinogens, which can often be shown to be mutagenic in vitro. The most notable example of this is lung cancer, which is aetiologically linked with the action of activated polycyclic hydrocarbons. As epoxides, these form adducts with DNA, which cannot be adequately repaired and thereby introduce mutations.

However, the conversion of a normal cell to a malignant cancer never occurs in a single step and cannot be attributed solely to the mutation of a single gene. Rather, there must be a series of changes in the properties of the collection of cells that make up the developing tumour,[2,3] and the evolution of a tumour towards an ever more malignant phenotype is a common clinical experience. The behaviour and severity of any cancer are therefore decided by a multifactorial range of genetic changes. As cancer is a disease of proliferating cells, it was not surprising to find that many of these mutations affect genes that control the rate of cellular proliferation and the ability of a cell to differentiate.

The oncogenes

It was believed initially that the genetic mutations responsible for cancer caused a deletion of essential regulatory genes restraining cell growth. However, the discovery of retroviruses radically changed these ideas on the alterations that occur to genes in cancer, led to the discovery of oncogenes, and gave a new perspective of cancer as being caused by genes that actively promote uncontrolled growth. In 1911, Peyton Rous described a transmissible agent that could pass tumours between chickens. It was subsequently shown that this agent was a retrovirus that harboured a specific gene, v-src, which was itself responsible for the tumours. More retroviruses were discovered that could transform cells to rapid and uncontrolled growth when grown in culture, and these too contained genes implicated in carcinogenesis. The term "oncogene" was coined to denote those viral genes that might have a role in converting normal cells to cancerous ones, and the action of the viral proteins was envisaged as helping to push the cell into a proliferative (cancerous) cycle.

In 1976, sequences of DNA related to the retroviral oncogene *v-src* were shown to be present in the DNA of normal uninfected chicken cells.[4] The cellular homologues of other, *v-onc*, sequences were identified subsequently. Hence it was recognised that some of the genes linked to cancer exist within normal cells before infection by retroviruses occurs. Elucidation of the life cycle of the retroviruses showed that they can infrequently hijack—transduce—incomplete portions of cellular genes into their own genetic material, which leads to the damage of the normal genes in such a way that they no longer function correctly. The cellular gene may either be placed under virally determined transcriptional control (both quantitatively and temporally), or sustain critical mutations to the coding sequence such that the function of the protein product is altered, or both.[2] If the transduced gene plays a central part in controlling growth and differentiation, these changes in structure and expression may contribute to the transformation of an infected cell.

Most non-viral oncogenes seem to be altered forms of cellular genes which encode proteins that participate in the pathways of cellular proliferation. The normal, intact cellular genes are known as proto-oncogenes. Cells receive signals to proliferate via growth factors that bind receptors located on the outside surface of the cell. The signals are then transmitted into the cell and across the cytoplasm to the nucleus. There, by transcriptionally activating the genes for proliferation and by suppressing the genes required for growth arrest or cellular differentiation, the signals are converted into growth responses. As a general rule, oncogenes are mutated forms of the cellular proto-oncogenes which encode the components of this signalling pathway (fig 14.1).[5] Moreover, the oncogenes are often changed relative to their proto-oncogene forebears in such a way that the proliferative signals are jammed in the "on" position.

Functions of cellular proto-oncogene products

Thus cellular proto-oncogenes can be broadly classified according to the cellular compartment in which their encoded

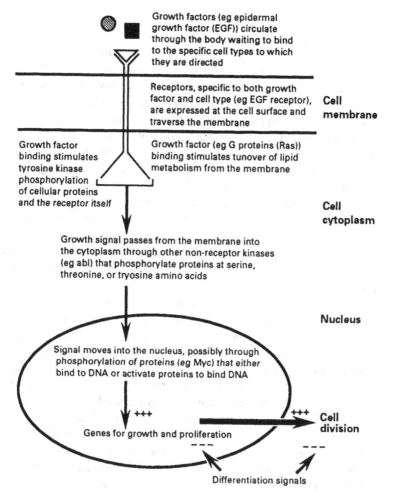

FIG 14.1—*Components of the cell proliferation signal pathway*

proteins are active. Several proto-oncogenes that encode growth factors (such as the platelet derived growth factor (PDGF), *c-sis*) are known.[6] Clearly, the inappropriate expression of such factors could lead to a continual growth signal being transmitted both to the cell that produces it and to that cell's close neighbours. In some instances the constitutive expression of these growth factors is thought to promote the autocrine growth of the cancer cell, possibly by the growth factor's association with the receptor within the cell (such as in the endoplasmic reticulum). The complex of growth factor and receptor then signals to the cell's own nucleus to maintain the proliferative signal.

Similarly, the products of several proto-oncogenes normally serve as growth factor receptors in the cell membrane (such as epidermal growth factor (EGF) receptor, erb-B). These molecules bind the growth factor and pass the growth signal to intracellular molecules, usually by activating one or more of the second messenger pathways involving inositol lipid turnover and protein kinases A and C. Although the precise signalling pathway is not clear, these receptors often possess tyrosine kinase activity (that is, after binding with growth factor they are capable of phosphorylating certain target molecules on tyrosine residues). Interestingly, the relevant cellular molecules that are the targets for this tyrosine phosphorylation are not known, but in the activated oncogenic form of the protein this enzymatic activity is usually increased and deregulated, which suggests that it is an important signalling property of growth factor receptors.[7]

Intracellular molecules on the inner side of the cytoplasmic membrane (such as the ras gene family) are then involved in propagating the growth signal, although again the precise biochemical mechanisms are not clear.[8][9] Other proto-oncogenes encode proteins, such as c-mos, that operate in the cytoplasm and transmit the signal from the membrane to the nucleus. Some of these proteins also possess kinase activity directed towards serine, threonine, or tyrosine residues, and deregulation of these activities is thought to contribute to the tumorigenic evolution of the cell.

Finally, many proto-oncogene products are located in the cell nucleus, which is where control of gene expression is principally regulated and so is the ultimate target of the growth signal. These proto-oncogene products (such as c-myc and molecules related to thyroid hormone receptor, such as erb-A) are usually involved directly in controlling gene transcription.[10]

In theory, any molecule that is involved in the pathways of cellular transmission of growth activating signals is a potential target for transforming mutations and may eventually be identified as a proto-oncogene. Many new oncogenes have now been identified both from sequences carried by acutely transforming retroviruses and by other techniques, and several are commonly detected in their mutated forms in human cancer cells. Table 14.1 gives a representative list of some oncogenes associated with transformation, and their grouping by cellular localisation.

TABLE 14.1—*Examples (not exhaustive list) of dominantly acting oncogenes that are activated by overexpression or mutation—to illustrate the classes of oncogenes (growth factors, membrane receptors, non-receptor protein kinases, and nuclear oncogenes) that have been identified from tumour cells*

Oncogene	Function of normal protein	Activation by
PDGF B chain (sis)	Growth factor for cells of mesenchymal origin	Originally isolated from a transforming retrovirus, SSV, fused to viral envelope gene. PDGF is released from many tumour cells, possibly contributing to transformation by autocrine stimulation of growth (megakaryoblastic leukaemia and breast cancer). Activation is generally by overexpression rather than mutation
hst int-2	Growth factors related to fibroblast growth factor	hst was isolated from DNA from human stomach cancer cells and is amplified in some glioblastomas and Kaposi's sarcoma; int-2 is amplified in subset of human breast cancers and glioblastomas. Both proto-oncogenes are angiogenic and may lead to neovascularisation of tumours
erb-1 erb-2 (neu)	EGF receptor (membrane tyrosine kinase receptor) and receptor kinase similar to EGF receptor (neu)	Transmembrane receptor protein can be activated in vitro by gross deletion (of extracytoplasmic domain) and point mutations, which usually deregulate the tyrosine kinase activity. In human tumours activation is principally by amplification (overexpression): neu is amplified in subsets of breast carcinomas; erb-1 is activated in breast, bladder, brain, and ovarian carcinomas, melanomas, and gliomas
fms kit ros	CSF receptor tyrosine kinase. Tyrosine kinase similar to PDGF receptor Insulin receptor tyrosine kinase	All transmembrane receptors with tyrosine kinase activities, which can become deregulated in activated molecules recovered from transformed cells in vitro. Activation is by mutations or deletions: ros is amplified in some breast carcinomas
abl	Protein located on inner side of cytoplasmic membrane; has tyrosine kinase activity	Activated by chromosomal translocation in CGL (see fig 14.2) as part of Philadelphia chromosome [t(9:22) (q34:q11)] in which tyrosine kinase domain is fused to bcr gene. Biochemical target of normal protein or mutated hybrid not known
pp60-src	Inner membrane, tyrosine kinase protein	First known retrovirally transduced oncogene. Normal function not known, but tyrosine kinase activity deregulated in squamous cell and stomach carcinomas

Table continued overleaf

TABLE 14.1—*Oncogenes activated by overexpression or mutation (continued)*

Oncogene	Function of normal protein	Activation by
ras	Family of G proteins of unknown function located at inner side of cytoplasmic membrane	Point mutations to amino acid positions 12, 13, or 61 correlate with activating mutations and the biochemical properties (see text). Activated in various leukaemias, colorectal, stomach, thyroid, bladder, lung, and ovarian carcinomas, neuroblastomas, and hepatomas
gsp	Inner membrane G protein	Activated in pituitary cancers
mil, mos pim	Serine/threonine protein kinases located in the cell cytoplasm	Activating point mutations increase kinase activities. Relevance to human cancers not yet clear
myc	Family of nuclear proteins (c-, N-, L-, B-myc). Transcription factors	Various *myc* genes amplified in several human tumours (such as breast, cervical, and lung carcinomas), which correlates with poor prognosis. Translocation of *c-myc* to an Ig locus (chromosome 2, 14, or 22) characteristic of Burkitt's lymphoma. Activation probably by inappropriate levels and timing of expression. High levels of *myc* genes may be incompatible with process of differentiation
fos, jun	Nuclear proteins that combine as heterodimers to form the transcription factor, AP-1	AP-1 regulates expression of genes of differentiation and may control DNA replication. Relevance to human cancers not yet clear
erb-A	"Zinc finger" containing nuclear protein: a thyroid hormone receptor and transcription factor	Binds specific DNA sequences in absence of hormone to repress transcription, which is removed when hormone present. Activating mutations (in vitro) remove ability of hormone to bind receptor, so repression of transcription (and differentiation) is not reversible
myb	Nuclear transcription factor	Amplified in several human tumours including colorectal and gastric carcinoma
Viral oncogenes		
p40 tax	HTLV, transcriptional activator	Activates transcription of viral and cellular genes (IL2 and IL2 receptor and other growth related genes). Associated with ATL
E6, E7	HPV types 16 and 18 nuclear proteins regulating HPV genome expression	Nuclear proteins that sequester cellular tumour suppressor genes (such as *rb* and *p53*) thereby removing critical differentiation signals. Associated with anogenital cancers

TABLE 14.1—*Oncogenes activated by overexpression or mutation (continued)*

Oncogene	Function of normal protein	Activation by
Others:		
CD44	Cell adhesion molecule	Variant spliced forms can dominantly confer metastatic capacity to non-metastatic cells in an animal model

PDGF = platelet derived growth factor; SSV = simian sarcoma virus; EGF = epidermal growth factor; CSF = colony stimulating factor; CGL = chronic granulocytic leukaemia; Ig = immunoglobulin; HTLV = human T cell leukaemia/lymphoma virus; IL = interleukin; ATL = adult T cell leukaemia; HPV = human papillomavirus.

Mechanisms of transformation of potential viral oncogenes by hepatitis B virus (HBV) (hepatocellular carcinoma) and Epstein–Barr virus (EBV) (Burkitt's lymphoma and nasopharyngeal carcinoma) not clearly defined.

Activation of proto-oncogenes to oncogenes

The result of proto-oncogene activation is the increased expression of genes signalling proliferation of cells that should not normally be proliferating. Direct comparisons of the DNA sequences of normal proto-oncogenes with their mutated oncogene derivatives show how proto-oncogenes become activated (that is, converted to the form in which their protein products encode proteins that participate in cell transformation). Activation can occur by changes to the coding sequence of the proto-oncogene so that a mutated protein with aberrant biochemical properties—an oncoprotein—is produced. Such activation events can occur by deletions of portions of the coding sequence (often to remove areas that are required for normal, negative control of the signalling function); by fusion of the sequence to other protein domains that alter the activity of the signalling domain (as occurs in several fusion proteins induced by chromosomal translocation such as the Philadelphia positive translocation in chronic granulocytic leukaemia (CGL), see fig 14.2); or by point mutations of the sequence at essential bases, which critically alter the amino acid sequence of the protein product. The *ras* class of proto-oncogenes can become oncogenic by as little as the change of a single base in the sequence of the gene. This alters the protein sequence by only a single amino acid. The Ras protein action is turned off by hydrolysis of guanosine triphosphate (GTP) to guanosine diphosphate (GDP). The single mutation in the oncogenic form

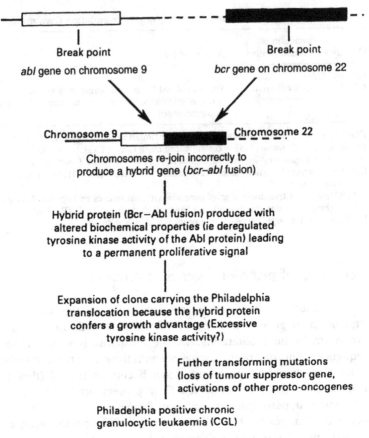

Break point
abl gene on chromosome 9

Break point
bcr gene on chromosome 22

Chromosome 9   Chromosome 22

Chromosomes re-join incorrectly to
produce a hybrid gene (bcr–abl fusion)

Hybrid protein (Bcr–Abl fusion) produced with
altered biochemical properties (ie deregulated
tyrosine kinase activity of the Abl protein) leading
to a permanent proliferative signal

Expansion of clone carrying the Philadelphia
translocation because the hybrid protein
confers a growth advantage (Excessive
tyrosine kinase activity?)

Further transforming mutations
(loss of tumour suppressor gene,
activations of other proto-oncogenes

Philadelphia positive chronic
granulocytic leukaemia (CGL)

FIG 14.2—*Activation of the* abl *proto-oncogene by chromosomal translocation (the Philadelphia translocation) to produce a hybrid Bcr–Abl fusion protein.*

of the Ras protein reduces the ability of the protein to hydrolyse guanosine triphosphate, so the signalling activity of the protein is continually turned on. Up to 15% of human cancers may have this variety of mutation of a *ras* gene.

Alternatively, activation can occur by mutation that deregulates the levels or timing, or both, of expressing the structurally unaltered proto-oncogene. For example, in the human cancer Burkitt's lymphoma, a chromosomal translocation causes the *myc* gene to become incorrectly linked to a chromosomal region that permits its permanent expression—even at the incorrect time, when the proliferative signal given to the cell by the Myc protein should be silent. In other human tumours certain proto-oncogenes become amplified, so many DNA copies of the gene exist within the cancer cell, instead of the normal two copies.

Generally it is chemical or physical carcinogens, such as ionising radiation, or mutagenic chemicals that act on DNA that induce mistakes in the cell's DNA replication machinery or chromosomal instability, which leads to some of the effects on gene structure described above. Infection by viruses, however, may also contribute to the genome instability that characterises the cancer cell and predisposes to genetic damage of this type. Such activation effects can allow proliferative signals to be transmitted, and these mask or compete out the normal signals for differentiation or growth inhibition. The application of molecular biology techniques has now shown that a relatively limited set of cellular genes is repeatedly involved in tumours which are caused by disparate carcinogens.

Interaction of oncogenes

In general, it seems that fully transformed cells must express at least two activated oncogenes, as well as other genetic abnormalities. Many human cancers that have a viral aetiology (about 20%) will also involve interactions of the viral proteins with a subset of these cellular proto-oncogenes. Such is the case with DNA tumour viruses (such as SV40 or papillomaviruses types 16 or 18), whereby the viral "oncogene" (which has not arisen by mutation of a normal cellular gene in these cases) can contribute to the transformation of a cell in much the same way as an activated cellular proto-oncogene. Fully transformed cells tend to express oncogenes that cooperate with each other,[11] and two classes of cooperating oncogenes can be broadly distinguished. At least one of those oncogenes that encode proteins of the membrane receptor or cytoplasmic class is usually expressed in tandem with an oncogene that expresses a protein that is found in the nucleus (table 14.1). Expression of a member of either class alone can often produce cellular immortalisation (the ability of a cell to grow indefinitely in culture) but not full tumorigenicity (the ability of a cell to form tumours when transplanted into animals). Each class of oncogene deranges the growth properties of the cell in subtly different ways, and the combination of effects is required for transformation of a cell to full tumorigenicity.

The tumour suppressor genes

Recent discoveries have begun to confirm earlier ideas that mutations in carcinogenesis also involve genes that negatively regulate cell proliferation (and, by inference, promote cell differentiation). Activated oncogenes are generally considered to be "dominantly acting" within the cell because mutation, even in a single copy, appears to be sufficient to drive the cell towards malignancy. However, cancer clearly can no longer be explained solely in terms of activated oncogene expression and mutation, because members of a second class of cellular genes are now emerging as being equally important in driving progression to malignant transformation.

Early cytogenic analysis showed that many human solid tumours show a reproducible loss of genetic material at defined chromosomal loci. These losses are specific for a given tumour type but differ between tumours. The absence of functional genes at these loci seemed to be related to tumour development and, moreover, both alleles (copies on homologous chromosomes) were sometimes seen to be lost (by deletion or gross chromosomal rearrangement). Through observations of the genetic epidemiology of the rare childhood cancer, retinoblastoma, Knudson suggested that a heterozygous child might inherit a (recessive) mutation to one allele of a gene that predisposes to tumours (an inherited first genetic "hit"). Such individuals will go on to develop cancer if the mutation is made homozygous by subsequent damage to the remaining intact allele (the second somatic genetic "hit").[12] This suggestion of a new class of cellular genes that regulate cell growth by counteracting the action of the proto-oncogenes led to the proposal that there are genes that are active at embryogenesis for proliferation ("transforming" genes), which are suppressed at differentiation by dominant "suppressor" genes.[13]

Knudson therefore named this proposed new class of genes "anti-oncogenes" (or "tumour suppressor genes"). The statistical analysis of the occurrence of retinoblastoma in affected families compared with the general population suggested that two "hits" are required, one in each allele of the suppressor gene, before a cell becomes fully predisposed to neoplastic growth. Cancer causing mutations in these genes can therefore be regarded formally as being recessive because both copies of the gene must be defective.

Tumour suppressor genes can be rationalised in terms of the conflicting needs for stem cells both to proliferate and to differentiate. Within a stem cell compartment, cycling cells will only differentiate in response to particular genetic or epigenetic signals; one class of genes (the proto-oncogenes) can simplistically be viewed as the genes required to keep the cells cycling, and the second class (the tumour suppressors) would be those genes that override the proliferative signals and induce the cell to differentiate. The loss of both copies of any of these suppressor genes, before the critical time in a cell's life history at which it is required to interpret the differentiation specific signals, would permit continued, inappropriate expression of the proliferative genes. Such a cell may then be launched on the first stages of neoplastic transformation.

Knudson's concept has been increasingly supported by the findings that several familial cancer syndromes are linked to the inherited loss of a defined chromosomal locus at one allele. This germinal mutation predisposes the individual to the cancer, but for the disease to emerge a second mutation is required somatically at the unaffected allele in the cell(s) that eventually become transformed. In cancers with a sporadic form as well as a familial form, the same chromosomal locus is affected in some cases, but both alleles are inactivated somatically. The differences in ages of patients, and features of the cancers, at presentation support such a model.

There is a rapidly growing list of tumours now being associated with homozygous loss of specific chromosomal loci.[14] Until genetic linkage studies have been carried out for each tumour type, however, it is not possible to know which losses represent germline alleles important in contributing a genetic predisposition to each specific cancer.

In addition, much in vitro research has implied the existence of genes capable of suppressing malignancy. For example, the fusion of a normal cell with a malignant cell produces a hybrid in which the tumorigenic phenotype is usually suppressed and the differentiation programme of the normal parent cell may be imposed upon the hybrid.[15] Subsequent chromosomal loss from the hybrid can occur, which leads to reappearance of tumorigenicity. The loss of specific chromosomal loci can be reproducibly associated with the reappearance of tumorigenicity, which suggests that tumour suppressive information resides at that locus. These and other experiments, whereby tumour cells can be shown to revert to the non-tumorigenic phenotype by transfer of specific pieces of DNA,[16] permit the identification of some of the proposed recessive loci that normally act to restrain cellular transformation.

The discovery of the tumour suppressor genes promises insights into the processes leading to normal differentiation and the way in which they may be disrupted during cellular transformation. Clinically, it provides a new opportunity to assess cancer susceptibility within families and to screen defined patients for certain types of cancer against which early intervention may be critical. These genes may also lead to new approaches to cancer treatment.

The proteins of some DNA tumour viruses that are known to be involved in causing cancer (particularly SV40 and several papillomaviruses) have been shown to bind to cellular tumour suppressor proteins such as p53 or the retinoblastoma (Rb) protein.[17] Hence the transforming virus proteins may well be acting by inactivating the Rb or p53 proteins. Infection of a cell before the implementation of the normal differentiation signals therefore removes the effects of the *rb* gene, which is probably central to the interpretation of those differentiating signals. Differentiation in the relevant compartment of cells is prevented and the infected cell misses its only opportunity to leave the proliferative cycle. It now becomes vulnerable to the action of any oncogenes that have already, or will later, become activated.

Several other tumour suppressor genes have now been identified, such as the Wilms' tumour (*WT*) gene, the gene deleted in cancer cells from patients with neurofibromatosis (*NF*), and the "deleted in colon carcinoma" (*DCC*) gene.[18] The neurofibromatosis gene has homology with the guanosine triphosphatase (GTPase) activating protein (GAP) which interacts with the Ras class of proto-oncogene proteins, thereby providing a mechanistic link between the tumour suppressors and oncogenes. In contrast, the *p53*, *rb*, and *WT* genes all encode transcription factors capable of regulating gene expression.

Loss of tumour suppressor gene expression may be a prerequisite for the development of most cancers, and mutation to the *p53* gene may actually represent the most common mutation in all human cancers.[19] For example, a heritable mutation in *p53* may be the "first hit" mutation in families with the Li–Fraumeni syndrome, whose members show a greatly increased risk of developing a range of tumours, including breast cancer, osteosarcoma, leukaemia, and soft tissue sarcoma. The inference is that p53 protein may mediate critical cycle decisions in a wide range of tissue types; the absence of the correct protein at certain times during the differentiation programme allows the cells to continue proliferating instead of differentiating.

The genes of metastasis

Although the genes controlling proliferation and differentiation are intuitively the most likely targets for oncogenic mutations, other genetic components of the progression to tumorigenicity are at least as important as the oncogenes and tumour suppressors, at least in terms of clinical progression and outcome of disease. Often, the localised primary tumour cell population itself may pose relatively little direct threat to the patient. Clinically, however, the most life threatening aspect of tumour cell growth is the ability of individual cells to spread from the site of primary growth to distant regions of the body where they can initiate secondary tumours, a process called metastasis. Metastasis is particularly dangerous to the patient because it increases the scope of damage that can be done by the rapidly growing tumour cell population and prevents that damage being restricted to a single site in the body.

Despite the overriding clinical importance of metastasis, the underlying mechanisms are only just becoming tractable to molecular study. Not all tumour cells have the same capacity to metastasise from within the same tumour population. Consequently, the progression from the partially transformed state (essentially a benign tumour mass) to the metastatic state is often considered to involve distinct, but interdependent, events. For tumour cells to become fully metastatic they must be able to acquire several properties that are not present in transformed but non-metastatic cells, and there is now considerable evidence that some or all of these processes are under genetic control.[20] Indeed, whereas it can take many years for a tumour to grow in situ, the conversion to malignancy can often occur in months in animal models, which suggests that relatively few events are required. In non-malignant tumorigenic rat carcinoma cells the only requirement for acquiring a fully metastatic phenotype has been shown to be the expression of a variant of the CD44 cellular adhesion molecule on the cell surface.[21] Similarly, increased expression of the nm23 gene in a melanoma cell line seems to suppress the malignant phenotype of these cells.[22] These studies suggest that the difference between tumorigenic but non-metastatic compared with fully metastatic behaviour may be attributable to the expression of only one or a few genes that promote metastasis or to the loss of an equally small number of genes that exert a suppressive effect on tumour metastasis. In the conversion from normal to tumour to metastatic cell, therefore, there are probably certain rate limiting steps in the overall sequence, which must occur before the malignant phenotype can evolve. For each tumour type these rate limiting steps may differ; for each metastasising system the individual rate limiting steps may depend on the tumour cell type, its requirements for growth factors, and its ability to escape the host immune response, especially when the tumour cells are exposed in the distributing system (that is, the blood).

The cancer cell and the immune system

Far less well understood, but of great potential importance in treatment, are the genes that control the reaction of cancer cells with the immune systems of patients. As cancer cells develop by a series of genetic and epigenetic mutations of genes within the once normal cells of the body, natural in vivo immune responses to cancer cells would in theory have to be autoimmune, as cancer cells might be expected to constitute an immunologically "self" population that the host's own immune system cannot recognise as foreign and therefore reject. In some circumstances this is assumed to be the case, and tumours can develop within the body unchecked by any natural immune reaction to the aberrant cells.

However, several lines of evidence suggest that cancer cells, far from representing an immunologically hidden population, actually express determinants—tumour antigens—that can form the basis of an anti-tumour immune reaction. Clinical evidence for specific anti-tumour immune activity is seen on the rare occasions when patients experience spontaneous disappearance of cancers, which cannot be attributed to the treatment, and the concept of anti-tumour "immune surveillance" may provide a partial explanation for the observed age related incidence of cancers and for the greatly increased incidence of various forms of cancers in patients whose immunity is suppressed by viral infection or other means. Transplant rejection may represent an in vivo defence mechanism against tumour cells that express antigens which seem to be foreign (as a result of activating oncogenic mutation).[23]

Largely unsuccessful attempts to immunise cancer patients with their own cancer cells are being replaced by treatment protocols aimed at enhancing the immune system's recognition and destruction of emerging solid tumours. These efforts now provide a focus for therapeutic intervention[24][25] and have stimulated the first in vivo trials for human gene treatment for cancer.[26] Antigen expression can be upregulated by cytokines, and the activity of the anti-tumour effector cells of the immune system can be increased by treating the patient with lymphokines or preferably by delivering cytokine genes to the site of the tumour, where their expression will promote a vigorous immune response by circulating T cells and monocytes (adoptive immunotherapy).[24] This is the theory behind clinical trials in which the interleukin-2 gene was transferred into tumour infiltrating lymphocytes (TIL), which were reinjected into patients with end stage melanoma. The presence of the interleukin-2 protein should activate these lymphocytes, which presumably have targeting specificity for the tumour (from which they were initially recovered). The early results, admittedly in melanomas that have proved refractory to all other treatments, are encouraging but far from spectacular.[26] (For further details, see chapter 20.)

The emergence of a tumour probably represents the outcome of a finely balanced immune surveillance system in which the rate of growth of the transformed cells has finally outstripped the capacity of the immune system to control it. Indeed, some of the critical mutations that finally produce the fully malignant phenotype may be of genes that shift the balance towards evasion of the immune response which otherwise has been successful in clearing, or controlling, the premalignant cells.

One specific example of a tumour antigen with immediate clinical importance has been P-glycoprotein (p170), the expression of which is associated with the phenotype giving resistance to many drugs.[27] The expression of this protein in tumour cells has dramatically severe clinical effects because various cytotoxic drugs used to target rapidly growing cells in patients with cancer are bound by p170 and subsequently transported out of the cell. P-glycoprotein is a membrane protein that acts as a pump dependent on ATP for the efflux of drugs and is often responsible for the disappointing recurrence of tumours after initially favourable responses to chemotherapy. In one study 85% of patients with primary, advanced breast carcinoma expressed P-glycoprotein in at least some of the tumour cells.[28] Such studies suggest that monitoring for expression of P-glycoprotein may help predict chemotherapy treatment failure and indicate changing the treatment regimen. Eventually, co-administering P-glycoprotein inhibitors may allow highly effective chemotherapy without the development of the phenotype that gives resistance to many drugs.

From laboratory to clinic

The development of a tumour cell population is a multifactorial process. There is rarely, if ever, a single event to which the conversion of a normal cell to a tumour cell can be solely attributed. Thus, a fully malignant cancer comprises a population of essentially similar cells that harbour many different genetic abnormalities compared with their normal counterparts. It is not clear, however, whether these mutations must occur in a specific temporal sequence.

Diagnosis and prognosis

The genetic changes that occur during cellular carcinogenesis are associated with the cellular acquisition of more aggressive growth characteristics and increasingly malignant behaviour. Specific mutations may, however, be essential for tumour types to emerge in different tissues, the normal differentiation programmes of which are determined by different regulatory genes (for example, mutation of the *rb* gene seems to be critical to allow cells of the retinal lineage to escape the normal rigours of growth control). Each tumour type may therefore require tissue specific, rate limiting, cellular mutations (tumour initiation) before it can evolve,[29] whereas the nature, timing, and combination of other mutations in the tumour cells may not be crucial (tumour progression).[30]

A main potential benefit of this molecular knowledge will therefore be an increasingly accurate assessment of risk of developing certain types of cancer. Familial cancer syndromes will become amenable to genetic diagnosis using restriction fragment length polymorphisms (RFLPs) of markers linked to the genes that are known to predispose to a given disease (see chapter 16). The findings that the loss of certain tumour suppressor genes leads to a high risk of developing cancer will permit the presymptomatic diagnosis of individuals at risk, thereby indicating intensive surveillance and pre-emptive intervention before the cancer develops. Retinoblastoma, Wilms' tumour, neurofibromatosis, and

Li–Fraumeni syndrome are some of the cancers the risk of which can now be reliably assessed in members of affected families. Similar studies should have begun to identify genes predisposing to more common cancers, such as heritable breast cancers.[31]

A knowledge of the specific oncogene activation events within a cancer cell population is also becoming an increasingly reliable diagnostic and prognostic indicator for disease progression. For example, the Philadelphia translocation can be used as a diagnostic marker for chronic granulocytic disease (this chromosomal translocation is found in about 90% of all patients with this disease). Similarly, patients presenting with acute lymphoblastic leukaemia (ALL) who have t(4;11) karyotypic abnormalities have a very poor clinical prognosis. Conversely, the clinician can now place patients diagnosed as having acute myeloid leukaemia in the best prognostic category if they present with abnormalities of chromosome 16 in the leukaemic cells. In the case of solid tumours, breast cancers that have lost expression of the *nm23* gene are strongly associated with very poor patient survival, and would indicate to the clinician a high probability of spread of the cancer cells to the axillary lymph nodes and other widespread metastasis. Finally, knowledge of the "active" oncogenes can be used to detect residual disease in patients—for example, patients with chronic granulocytic leukaemia who are in remission can be screened with the polymerase chain reaction (PCR) for leukaemic cells containing the *bcr-abl* gene. Early detection of relapse by molecular means, before disease has clinically reappeared, will permit better management of patients with recurring disease.

Hopes for specific treatments targeted at cancer cells

The ultimate goal of this wide range of research must therefore be to lead to mechanism based treatments for the disease, and yet many years of research have still not made a large impact on current treatments. The individual stages at which intervention into the molecular pathways of cancer cell proliferation might occur are manifold but also, for that reason, problematic. Pharmaceutical blocking of any of the oncogene products active in a cancer, using rational drug design against the structures (from crystallographic structural studies) known to be altered in the oncoprotein, is therefore attractive; alternatively, it may prove possible to deliver to cancer cells anti-sense DNA or RNA constructs, which block expression of the activated form of an oncogene while leaving the unmutated gene unaffected. In addition, monoclonal antibodies targeted specifically at the mutated regions of an oncoprotein may be able to differentiate between oncogene products and the normal cellular equivalents, which would lead to clearance of cancer cells but not normal cells. Quite apart from the technical difficulties of targeted delivery of these therapeutic agents in the patient, however, by the time the malignant phenotype has emerged the tumour cell may no longer depend on continued expression of any individual mutant protein. Likewise, replacing the functions provided by genes that have been lost in tumour cells—the tumour suppressor genes—has been proposed as a means of reverting to the transformed phenotype (gene therapy). But the loss of such genes may be an ancient event, relative to the growth requirements of the cancer cell, by the time the clinician first sees the patient. Additionally, the interaction of the cancer population with the local and distant environment is central to understanding the life threatening aspects of metastasis. Finally, much poorly understood immunology underlies the control of the emerging cancer cell population by the host immune system. Mutations that lead to loss of the immunological balance between tumour cell destruction and immune escape, and the factors required to restore that balance in favour of tumour cell clearance, are only now beginning to be addressed in molecular terms.

Currently, treatment for cancer can often be as distressing as the disease it seeks to cure. Recent research has shown that primary and secondary tumour populations consist of cells that have undergone multiple mutations, the molecular bases of which are becoming increasingly understood. It is to be hoped that these defects can be exploited by clinicians to target individual, or combinations of, critical cellular mutations, which will allow selective tumour cell killing leaving the remaining normal cells, and therefore the patient, largely untouched by the treatment.

1 Fischer J, Boveri T. *The origin of malignant tumours*. Boveri M, transl. Baltimore: Williams and Wilkins, 1929. Boveri T. *Zur Frage der Erstehung Maligner Tumoren*. 1914.
2 Bishop JM. Molecular themes in oncogenesis. *Cell* 1991;64:235–48.
3 Bourne HR, Varmus HE (eds). Oncogenes and cell proliferation. *Curr Opin Gen Devel* 1992;2:227–35.
4 Stehelin D, Varmus HE, Bishop JM, Vogt PK. DNA related to the transforming genes of avian sarcoma viruses is present in normal avian DNA. *Nature* 1976; 260:170–3.
5 Cantley LC, Auger KR, Carpenter C, Duckworth B, Graziani A, Kapeller R, et al. Oncogenes and signal transduction. *Cell* 1991;64:281–302.

6 Cross M, Dexter TM. Growth factors in development, transformation and tumorigenesis. *Cell* 1991;64:271–80.

7 Ullrich A, Schlessinger J. Signal transduction by receptors with tyrosine kinase activity. *Cell* 1990;61:203–12.

8 Bourne HR, Sanders DA, McCormick F. The GTPase superfamily: a conserved switch for diverse cell functions. *Nature* 1990;348:125–32.

9 Bourne HR, Sanders DA, McCormick F. The GTPase superfamily: conserved structure molecular mechanism. *Nature* 1990;349:117–27.

10 Forrest D, Curran T. Crossed signals: oncogene transcription factors. *Curr Opin Genet Devel* 1992;2:19–27.

11 Hunter T. Cooperation between oncogenes. *Cell* 1991;64:249–70.

12 Knudson AG. Mutation and cancer: statistical study of retinoblastoma. *Proc Natl Acad Sci USA* 1971;68:820–3.

13 Comings DE. A general theory of carcinogenesis. *Proc Natl Acad Sci USA* 1973; 70:3324–8.

14 Weinberg RA. Tumour suppressor genes. *Neuron* 1993;11:191–6.

15 Harris H. The analysis of malignancy by cell fusion: the position in 1988. *Cancer Res* 1988;48:3302–6.

16 Huang SH-J, Yee J-K, Shew J-Y, Chen PL, Bookstein R, Friedmann T, *et al.* Suppression of the neoplastic phenotype by replacement of the RB gene in human cancer cells. *Science* 1988;242:1563–6.

17 Levine AJ. The p53 protein and its interactions with the oncogene products of the small DNA tumor viruses. *Virology* 1990;177:419–26.

18 Marshall CJ. Tumour suppressor genes. *Cell* 1991;64:313–26.

19 Hollstein M, Sidransky D, Vogelsttein B, Harris CC. p53 mutations in human cancers. *Science* 1991;253:49–53.

20 Liotta LA, Steeg PS, Stetler-Stevenson WG. Cancer metastasis and angiogenesis: an imbalance of positive and negative regulation. *Cell* 1991;64:327–36.

21 Gunthert U, Hofman M, Rudy W, Reber S, Zoller M, Haussmann I, *et al.* A new variant of glycoprotein CD44 confers metastatic potential to rat carcinoma cells. *Cell* 1991;65:13–24.

22 Leone A, Flatow U, Richter King C, Sandeen MA, Margulies IMK, Liotta LA, *et al.* Reduced tumor incidence, metastatic potential, and cytokine responsiveness of nm23-transfected melanoma cells. *Cell* 1991;65:25–35.

23 Thomas L. In: Lawrence HS, ed. *Cellular and humoral aspects of the hypersensitivity states.* London: Cassell, 1959:529–31.

24 Russell SJ. Lymphokine gene therapy for cancer. *Immunol Today* 1990;11: 196–200.

25 Russell SJ. Interleukin-2 and T-cell malignancies: an autocrine loop with a twist. *Leukaemia* 1989;3:755–7.

26 Rosenberg SA, Abersold P, Cornetta A, Kasid A, Morgan RA, Moen, R, *et al.* Gene transfer into humans—immunotherapy of patients with advanced melanoma, using tumour infiltrating lymphocytes modified by retroviral gene transduction. *N Engl J Med* 1990;323:570–8.

27 Endicott JA, Ling V. The biochemistry of P-glycoprotein-mediated multidrug resistance. *Annu Rev Biochem* 1989;58:137–71.

28 Verrelle P, Meissonnier F, Fonck Y, Feiuel V, Dioneto G, Kwiatowski F, *et al.* Clinical relevance of immunohistochemical detection of multidrug resistance P-glycoprotein in breast carcinoma. *J Natl Cancer Inst* 1991;83:111–16.

29 Haber DA, Housman DE. Rate-limiting steps: the genetics of pediatric cancers. *Cell* 1991;64:5–8.

30 Stanbridge EJ. Identifying tumor suppressor genes in human colorectal cancer. *Science* 1990;247:12–13.

31 Miki Y *et al.* A strong candidate for the breast and ovarian cancer susceptibility gene BRCA1. *Science* 1994;266:66–71.

15: Human congenital malformations: insights from molecular genetics

Paul M Brickell

Human congenital malformation syndromes

It is estimated that 1·7–3·3% of live births are affected by congenital malformation,[1] and a very large number of syndromes have been identified.[2] The causes of the great majority of congenital malformations are unknown, but a recent survey[3] suggested that about 7·5% can currently be attributed to single gene defects, 6% to gross chromosomal abnormalities such as trisomy 21 in Down's syndrome, 20% to a combination of genetic and environmental factors, and 6·5% to maternal influences such as infection. To assist the study of human congenital malformation syndromes, and in particular to aid gene mapping, Winter and Baraitser established the microcomputer based London Dysmorphology Database.[2] This lists over 2000 malformation syndromes, excluding chromosomal syndromes such as Down's syndrome, and provides detailed descriptions of phenotypes and clinical photographs stored on CD-ROM. A subset of the database can be accessed at the UK Human Genome Mapping Project via the World Wide Web at http://www.hgmp.mrc.ac.uk/DHMHD/dysmorph.html, where it is linked to the Mouse Malformation Database[4] and a Mouse/Human Chromosome Homology Database.[5] This assists gene mapping by allowing investigators to search for human syndromes located at a chromosome region that is syntenic with a specific mouse chromosome region, and vice versa.

In recent years, something in the region of 200 genes for human congenital malformations have been identified and cloned, and

TABLE 15.1—*Some of the many congenital malformation syndromes in which a defective gene has been identified*

Condition	System(s) primarily affected	Defective *gene*/ gene product	Probable function of gene product	Reference
Aniridia	Eye	*PAX6*	Transcription factor	6
Hirschsprung's disease	Enteric nervous system	*RET*	Receptor protein-tyrosine kinase	7
Kallmann's syndrome	Olfactory lobes, gonads	*KAL*	Cell adhesion molecule	8
Lesch–Nyhan syndrome	Brain	*HGPRT*	Metabolic enzyme	9
Marfan's syndrome	Skeleton, eye, cardiovascular system	Fibrillin	Structural protein	10
Metaphyseal chondrodysplasia type Schmid	Long bones	Collagen X	Structural protein	11
Myotonic dystrophy	Muscle	DM kinase	Cytoplasmic protein–serine/threonine kinase	12
Neurofibromatosis type I	Multiple systems, neurofibromas	Neurofibromin	GTPase	13
Polycystic kidney disease (autosomal dominant type)	Kidney	*PKD1*	Cell adhesion molecule	14
Skeletal dysplasias	Skull, long bones	*FGFR1*, *FGFR2*, *FGFR3*	Receptor protein tyrosine kinases	15
Treacher Collins syndrome	Craniofacial structures	*Treacle*	Unknown	16
Waardenburg's syndrome type I	Hearing, pigmentation	*PAX3*	Transcription factor	17
Waardenburg's syndrome type II	Hearing, pigmentation	*MITF*	Transcription factor	18
X linked deafness	Hearing	*Brain 4*	Transcription factor	19

some of these are listed in table 15.1. In most cases, progress has resulted from the application of the positional cloning and candidate gene approaches that are described elsewhere in this book (see chapters 16 and 17). We are now witnessing the start of an era of fruitful collaboration between molecular geneticists and developmental biologists, as we seek to understand the normal functions of these genes in human embryonic development and to determine the effects of mutations on gene function. The purpose of this chapter is to illustrate the kind of progress that has been possible by reference to two types of human congenital malformation. These are aniridia, which affects the developing eye, and a group of syndromes that affect the developing skeleton, including achondroplasia, hypochondroplasia, thanatophoric dysplasia, and the craniosynostosis syndromes. Two quite different approaches were taken to determine the genetic bases of these two types of disorder, and a number of important lessons emerge from each.

Aniridia

Identification of PAX6 as the gene that is defective in aniridia

The main characteristic of aniridia is the almost complete absence of the iris,[6] although affected individuals may also have cataracts, corneal vascularisation, and glaucoma. In about two thirds of cases, aniridia is inherited as an autosomal dominant trait. The remaining cases are sporadic and may be associated with other abnormalities in the so called WAGR syndrome (predisposition to Wilms' tumour, aniridia, genitourinary abnormalities, mental retardation or handicap).

In the late 1970s, people with WAGR syndrome were found to have deletions involving chromosome band 11p13 and this important clue led eventually to the positional cloning of the aniridia gene.[20] During this procedure a mouse mutation called Small eye, which causes eye defects similar to those found in aniridia, was mapped to a region of mouse chromosome 2 that is syntenic to human 11p13, and this information was valuable in identifying the human aniridia gene.[6]

Nucleotide sequencing of the human aniridia gene showed that it was the homologue of a mouse gene named Pax6.[20] This gene had just been discovered independently on the basis of its possession of a sequence motif called a "paired box". This motif was originally discovered in a gene called paired, which is important for the correct development of the fruit fly Drosophila. At about the same time, mutation of the mouse Pax6 gene was shown to be responsible for the Small eye phenotype.[21] We now know that vertebrates have a family of at least nine Pax genes, named PAX1 to PAX9 in humans.

The proteins encoded by these genes have similar structures and appear to play similar roles in different parts of the developing embryo. Interestingly, at least two other PAX family members are also involved in human congenital malformations. Mutations in *PAX2* cause developmental abnormalities in the kidney and optic nerve,[22] whereas *PAX3* is mutated in Waardenburg's syndrome type I.[17] This condition is characterised by deafness and abnormal pigmentation, and is believed to result from abnormalities in the behaviour of neural crest cells.

The discovery of the link between aniridia and *PAX6* therefore illustrates a number of recurring themes in the study of human congenital malformations:

- the value of cytogenetic analysis of families with malformation syndromes
- the importance of positional cloning approaches for identifying disease genes
- the value of comparative studies with mouse malformation syndromes
- the lurking presence of *Drosophila* in studies of vertebrate development
- the idea that different members of a gene family, which play key roles in the development of different body parts, may be involved in different human disorders.

Structure and function of PAX6

As shown in figure 15.1, PAX6 protein contains two DNA binding domains, named the *paired* domain and the *paired* type homeodomain, and a carboxyl terminal PST domain (for

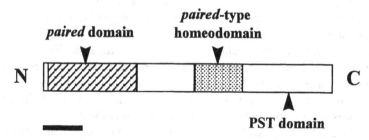

50 amino acids

FIG 15.1—*Sketch of the PAX6 protein, showing functional domains. N, amino terminus; C, carboxyl terminus.*

proline–serine, threonine rich) which acts as a transcriptional activator in vitro[23] (see chapter 13 for a discussion of transcriptional activation). PAX6 regulates its own expression[24][25] and there is evidence that it may also regulate transcription of genes encoding lens crystalline proteins[26] and the neural cell adhesion molecule L1.[27]

Mutations have now been defined in a large number of individuals with aniridia.[6] In all but one case, these mutations lead to premature termination of translation of PAX6 mRNA. In WAGR patients, the entire *PAX6* gene is deleted, along with neighbouring genes. It therefore seems likely that aniridia results from a complete loss of function of the mutated *PAX6* gene. As aniridia is an autosomal dominant trait, the developing iris must be particularly sensitive to the precise levels of PAX6 protein present. Other congenital eye malformations may result from partial loss of function of the mutated *PAX6* gene. For example, missense mutations resulting in amino acid substitutions in PAX6 have been found in individuals with abnormalities in the anterior chamber of the eye, whereas people with autosomal dominant keratitis, which is characterised by corneal opacification, corneal vascularisation, and foveal hypoplasia, have a mutation that results in a minor truncation of the carboxyl terminal end of PAX6.[28]

Homozygous *Small eye* mice, and a rare human individual who inherited two mutated *PAX6* genes, have no functional PAX6 and exhibit more extensive defects.[6] They completely lack eyes and nasal cavities, and have severe abnormalities in brain development. This indicates that PAX6 is involved in the development of structures other than the iris and cornea, but that these are less sensitive than the iris and cornea to partial reductions in PAX6 levels.

The pattern of *Pax6* expression in developing embryos, as determined by in situ hybridisation to *Pax6* gene transcripts (see chapter 2 for details of this method), is generally consistent with the pattern of abnormalities seen in *Pax6* mutants. Expression studies have largely been confined to animal models favoured by developmental biologists, such as mouse, chick, and zebra fish, although preliminary work indicates that the pattern of *PAX6* expression will prove to be similar in human embryos.[29] In normal mouse embryos, *Pax6* is expressed in a broad region of surface ectoderm covering the developing forebrain.[13] Expression is then generally downregulated but is maintained in the developing lens placode, which is a thickening of the surface ectoderm that will give rise to the lens and cornea. *Pax6* is also expressed in the underlying neuroepithelium, which will give rise to the retina, in specific regions of the developing brain, including the cerebellum, and in the olfactory neuroepithelium. Defects in each of these structures are associated with complete loss of *Pax6* function, as discussed above. Like the eye lens, the nasal cavities develop from ectodermal placodes, and so it has been suggested that Pax6 may be involved in the transition of presumptive lens ectoderm and presumptive nasal ectoderm to lens placode and nasal placode, respectively. The role of Pax6 in iris development is unclear. Studies of *Small eye* rats, which also carry a defective *Pax6* gene, have identified abnormalities in the migration of populations of neural crest cells which might normally give rise to iris stromal cells and, indeed, it has been suggested that these abnormalities could also underlie the failure of lens placode and nasal placode formation.[30]

Flies and the future

A recent and fascinating development in this story was the discovery that *Drosophila* has a gene with extensive homology to human *PAX6*.[31] This *PAX6* like gene, which is distinct from the drosophila *paired* gene, turned out to be the gene that is defective in a drosophila mutant called *eyeless*, in which there is a complete failure of eye development. Moreover, when the *eyeless/PAX6* like gene was expressed ectopically in regions of fly embryos that give rise to wings, legs, or antennae, flies developed with morphologically normal eyes located on these appendages.[32] These studies turn on its head the usual sequence of events, in which genes that are important in drosophila development are cloned and used as probes to search for vertebrate homologues, which then become candidates for roles in vertebrate development. Perhaps even more startling is the finding that the nematode worm *Caenorhabditis elegans* has a homologue of the *Pax6* gene that is required for development of its sense organ.[33]

It is staggering to think that development of vertebrate and insect eyes, and of the nematode sense organ, is under similar genetic control, in spite of the enormous differences in the morphology and mode of development of these structures. Recent work has, however, also revealed similarities between the genetic control of wing development in chickens and flies.[34] Indeed, these studies have suggested that whole genetic pathways may have been conserved through evolution,[35] and this raises the exciting possibility that the upstream regulators and downstream targets of human PAX6 may also be conserved in *Drosophila*. Identification of other components of the genetic pathway in which *eyeless* operates will therefore suggest candidates for upstream regulators and downstream targets of human PAX6.

Clues to the identities of other components of the PAX6 pathway may also come from the identification of genes involved in other congenital abnormalities of human eye development. For example, Gillespie's syndrome is an autosomal recessive trait characterised by partial aniridia, cerebellar ataxia, and mental handicap. Recent data show that *PAX6* is not involved in this condition, leaving open the possibility that it is caused by another gene in the PAX6 pathway.[36]

Congenital skeletal malformation: achondroplasia, hypochondroplasia, thanatophoric dysplasia, and the craniosynostosis syndromes

Recent work has shown that mutations in fibroblast growth factor receptor (*FGFR*) genes are responsible for a number of human congenital skeletal malformation syndromes.[15] These are listed in table 15.2.

Achondroplasia (ACH), hypochondroplasia (HCH), and thanatophoric dysplasia (TD) are primarily disorders of the long bones, although ribs and some craniofacial bones are also affected.[15] These bones develop by the packing together, or condensation, of undifferentiated mesenchyme cells in the embryo, followed by formation of cartilage by the condensed cells. The cartilage then acts as a template for deposition of bone by a process called endochondral ossification. Subsequent growth in the length of the bone occurs as a result of activity at the cartilaginous growth plate, located towards the ends of the bones. It is the growth plate that is affected in these conditions, resulting in shortened bones.[15]

In contrast, the craniosynostosis syndromes are primarily disorders of the flat bones of the skull.[15] These so called intramembranous bones develop directly from embryonic mesenchyme cells, without prior formation of cartilage. The malformations in the craniosynostosis syndromes result from premature fusion of the sutures between the skull bones, leading to malformation of the head. There are also malformations of the hands and/or feet in some cases.

TABLE 15.2—*Features of human congenital malformation syndromes involving FGFR genes*[15]

	Limb abnormalities	Craniofacial abnormalities	Other abnormalities	Mutated gene(s)
Disorders of long bones				
Achondroplasia (ACH)	Short	Large skull, under-developed mid-face	Spinal curvature, short base of skull	*FGFR3*
Hypochondro-plasia (HCH)	Short			*FGFR3*
Thanatophoric dysplasia (TD)	Short	Large skull, under-developed mid-face	Neonatal lethal, short ribs	*FGFR3*
Craniosynostosis syndromes				
Apert's syndrome (AS)	Abnormal fingers and toes	Craniosynostosis, under-developed mid-face	Severe	*FGFR2*
Crouzon's syndrome (CS)	Normal	Craniosynostosis, under-developed mid-face		*FGFR2*
Jackson–Weiss syndrome (JWS)	Normal fingers, abnormal toes	Craniosynostosis, under-developed mid-face		*FGFR2*
Pfeiffer's syndrome (PS)	Broad thumbs and big toes	Craniosynostosis, under-developed mid-face		*FGFR1* *FGFR2*

Craniosynostosis syndromes are characterised by premature fusion of the sutures between the flat bones of the skull, and are distinguished largely by differences in malformations of hands and feet. An under-developed mid-face can result from malformation of either the endochondral or the intramembranous bones of the head.

The syndromes listed in table 15.2 are all inherited as autosomal dominant conditions, although sporadic cases also occur as a result of new mutations.[15] Connections between these conditions and *FGFR* genes were made by a "candidate gene" approach, which contrasts with the "positional cloning" approach responsible for discovery of the association between *PAX6* and aniridia.

Fibroblast growth factors and their receptors

The story really begins over 10 years ago with the isolation of the fibroblast growth factors (FGFs) as serum factors capable of stimulating cell growth.[37] Nine structurally related FGFs (FGF-1 to FGF-9) have now been identified, and a number of *FGF* genes are known to be proto-oncogenes (see chapter 14). FGFs are secreted proteins that bind to extracellular matrix and act as local signals by binding to transmembrane receptors (FGFRs) on neighbouring cells.[15 37] Binding of an FGF to an FGFR induces dimerisation of the receptor and activation of its intrinsic protein tyrosine kinase activity. This triggers a sequence of intracellular events, one consequence of which is to alter the activities of transcription factors in the cell's nucleus and so alter the pattern of cellular gene expression (see chapter 9 for a discussion of signalling via transmembrane protein tyrosine kinase receptors). Four closely related FGFRs (FGFR-1 to FGFR-4) have been identified in humans,[15 37] and it is not yet entirely clear which FGFs bind to which FGFRs. The basic structure of the FGFR proteins is illustrated in figure 15.2, although each of the four *FGFR* genes actually encodes a range of variant proteins as a result of alternative splicing (see chapter 13).

Clues from developmental biologists: expression and activities of FGFs *and* FGFRs

Studies of *FGF* and *FGFR* gene expression in animals, of the effects of adding FGFs to vertebrate embryos, and of the consequences of abolishing FGFR expression in transgenic mice suggested a number of roles for FGF signalling during embryonic development. For example, FGFs appear to be important in controlling mesoderm induction, which is one of the earliest steps of tissue differentiation in the gastrulating embryo.[38] FGFR1 and FGFR2 may mediate FGF signals in this context, because transgenic mice lacking both copies of their *FGFR1* or *FGFR2* genes die early in embryonic development.[15] Later in development, FGFs have an important role in controlling outgrowth of early limb buds, and one of their most striking activities is to stimulate formation of a complete extra limb when "painted" on to the flank of a developing chick embryo.[39] Roles in development of the nervous system have also been suggested.[37] Among the in situ hybridization data gathered soon after the isolation of the *FGFR* genes were the findings that *FGFR1*, *FGFR2*, and *FGFR3* are co-expressed in regions of tissue that will go on to form cartilage and/or bone in the limbs, vertebral column, and skull,[40] whereas *FGFR3* is expressed in the growth plates of long bones during endochondral

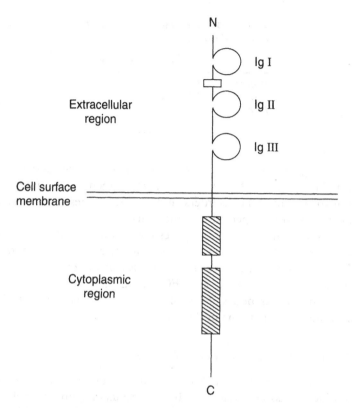

FIG 15.2—*Sketch of a fibroblast growth factor receptor (FGFR). IgI, IgII, and IgIII are the first, second, and third immunoglobulin like domains. N, amino terminus; C, carboxyl terminus; white box, acidic region; shaded boxes, bipartite protein tyrosine kinase domain interrupted by a "kinase insert" sequence.*

ossification.[41] Although these did not receive a great deal of attention at the time, they gained significance in light of the later discovery that *FGFR* genes were genetically linked to the genes responsible for skeletal dysplasias in humans.

Clues from molecular geneticists: chromosomal locations

Genetic linkage analysis allowed a number of the syndromes shown in table 15.2 to be mapped to defined chromosomal regions. Previous work in other laboratories had identified the chromosomal location of the *FGFR* genes (table 15.3), and inspection of published databases showed that these lay within the regions

TABLE 15.3—*Chromosomal location of human FGFR genes*[15]

Gene	Chromosomal location
FGFR1	8cen
FGFR2	10q25–q26
FGFR3	4p16
FGFR4	5q35

containing the disease genes. Knowledge of the embryonic expression patterns of the *FGFR* genes made them good candidates for involvement in the disorders, and so *FGFR* genes from affected individuals were inspected for mutations.

As shown in table 15.2, the results demonstrated that these disorders were indeed the result of mutations in *FGFR* genes. Moreover, they indicate roles for FGFR3 in the growth of long bones, and for FGFR1 and FGFR2 in the development of intramembranous bones. FGFR4 has not yet been found to be associated with any human disease.

The effects of mutation on FGFR function

A large number of different FGFR mutations have been identified.[15] Most of these lie in the extracellular region, particularly in the third immunoglobulin like domain (fig 15.2), and most involve amino acid substitutions. Individuals who have completely lost one copy of the *FGFR3* gene have normal skeletons,[42] as do transgenic mice that are heterozygous for *FGFR1*, *FGFR2*, or *FGFR3* gene deletions.[15] This strongly suggests that the autosomal dominant *FGFR* mutations found in human skeletal malformation syndromes do not abolish *FGFR* function, in contrast to the inactivation of *PAX6* function by mutations responsible for aniridia. One intriguing possibility is that the mutations lead to FGF independent dimerisation of the FGFRs, resulting in the constitutive activation of downstream signalling pathways.[43] This would resemble the constitutive activation that results from a point mutation in the extracellular region of the colony stimulating factor 1 receptor (CSF1R) protein tyrosine kinase, or from truncation of the extracellular region of the EGFR protein tyrosine kinase, although the effect of deregulating these two receptors is to provoke malignant transformation of cells (see chapter 14). Further biochemical studies of the naturally occurring mutant human

FGFRs, both in vitro and in transgenic mice engineered to express them, are likely to make a very important contribution to our understanding of signalling mediated by FGFRs and by receptor protein tyrosine kinase in general.

In support of the idea that the *FGFR* mutations activate FGFR function, transgenic mice that lack both copies of the *FGFR3* gene have overgrown long bones—the opposite phenotype to that resulting from *FGFR3* mutations in ACH, HCH, and TD.[15] This leaves us with the paradox that constitutive activation of a growth factor receptor results in inhibition of bone growth, whereas removal of the receptor stimulates growth. One possible explanation is that the normal role of FGFs in the growth plate is to induce cells to differentiate, and that activating mutations in *FGFR3* induce precocious differentiation and cessation of bone growth. A similar mechanism could account for the premature fusion of skull bone sutures in the craniosynostosis syndromes. Whatever the case, the discovery of these mutations is stimulating a great deal of interest among developmental biologists, and there are certain to be more surprises emerging from studies of the cellular basis of their effects on the growth plates of long bones and on the flat bones of the skull.

Conclusion

Recent advances in our ability to analyse gene function during embryonic development mean that we are very well placed to capitalise on the discovery of genes involved in human congenital malformations. In the years ahead we will come to understand the developmental mechanisms underlying a growing number of such abnormalities, and this will equally inform our understanding of normal embryonic development. Converting this information into improved treatment for those who have these distressing conditions will be a considerable challenge. Incidentally, in case you were wondering, FGFs have not yet been identified in fruit flies.

1 Weatherall DJ. *The new genetics in clinical practice*, 3rd edn. Oxford: Oxford University Press, 1991.

2 Winter RM, Baraitser M. *The London dysmorphology database*. Oxford: Oxford University Press, 1993.

3 Moore GE. Molecular genetic approaches to the study of human craniofacial dysmorphologies. *Int Rev Cytol* 1995;**158**:215–77.

4 Winter RM. A mouse malformation mutant supplement to the London Dysmorphology Database. *Am J Med Genet* 1988;**30**:812–19.

5 Searle AG, Edwards JH, Hall JG. Mouse homologues of human hereditary disease. *J Med Genet* 1994;**31**:1–19.

6 Hanson I, van Heyningen V. Pax6: more than meets the eye. *Trends Genet* 1995; **11**:268–72.

7 Pasini B, Ceccherini I, Romeo G. *RET* mutations in human disease. *Trends Genet* 1996;**12**:138–44.

8 del Castillo I, Cohen-Salmon M, Blanchard S, Lutfalla G, Petit C. Structure of the X-linked Kallmann syndrome gene and its homologous pseudogene on the Y chromosome. *Nat Genet* 1992;**2**:305–10.

9 Gibbs RA, Caskey CT. Identification and localization of mutations at the Lesch–Nyhan locus by ribonuclease A cleavage. *Science* 1987;**236**:303–5.

10 Karttunen L, Raghunath M, Lonnqvist L, Peltonen L. A compound–heterozygous Marfan patient: two defective fibrillin genes result in a lethal phenotype. *Am J Hum Genet* 1994;**55**:1083–91.

11 Olsen BR. Mutations in collagen genes resulting in metaphyseal and epiphyseal dysplasias. *Bone* 1995;**17**(suppl):45–9S.

12 Wang J, Pegoraro E, Menegazzo E, Gennarelli M, Hoop RC, Angelini C, *et al.* Myotonic dystrophy: evidence for a possible dominant-negative RNA mutation. *Hum Mol Genet* 1995;**4**:599–606.

13 Nakafuku M, Nagamine M, Ohtoshi A, Tanaka K, Toh-e A, Kaziro Y. Suppression of oncogenic Ras by mutant neurofibromatosis type I genes with single amino acid substitutions. *Proc Natl Acad Sci USA* 1993;**90**:6706–10.

14 Harris PC, Ward CJ, Peral B, Hughes J. Autosomal dominant polycystic kidney disease: molecular analysis. *Hum Mol Genet* 1995;**4**:1745–9.

15 Muenke M, Schell U. Fibroblast-growth-factor receptor mutations in human skeletal disorders. *Trends Genet* 1995;**11**:308–13.

16 The Treacher Collins Syndrome Collaborative Group. Positional cloning of a gene involved in the pathogenesis of Treacher Collins syndrome. *Nat Genet* 1996;**12**:130–6.

17 Tassabehji M, Newton VE, Liu XZ, Brady A, Donnai D, Krajewska-Walasek M, *et al.* The mutational spectrum in Waardenburg syndrome. *Hum Mol Genet* 1995;**4**:2131–7.

18 Moore KJ. Insight into the *microphthalmia* gene. *Trends Genet* 1995;**11**:442–8.

19 Cremers FP, Bitner-Glindzicz M, Pembrey ME, Ropers HH. Mapping and cloning hereditary deafness genes. *Curr Opin Genet Dev* 1995;**5**:371–5.

20 Ton CC, Hirvonen H, Miwa H, Weil MM, Monaghan P, Jordan T, *et al.* Positional cloning and characterization of a paired box- and homeobox-containing gene from the aniridia region. *Cell* 1991;**67**:1059–74.

21 Hill RE, Favor J, Hogan BL, Ton CC, Saunders GF, Hanson IM, *et al.* Mouse *small eye* results from mutations in a *paired*-like homeobox-containing gene. *Nature* 1991;**354**:522–5.

22 Sanyanusin P, Schimmenti LA, McNoe LA, Ward TA, Pierpoint ME, Suillivan MJ, *et al.* Mutation of the *PAX2* gene in a family with optic nerve colombas, renal anomalies and vesicoureteral reflux. *Nat Genet* 1995;**9**:358–63.

23 Glaser T, Jepeal L, Edwards JG, Young SR, Favor J, Maas RL. *PAX6* gene dosage effect in a family with congenital cataracts, aniridia, anophthalmia and central nervous system defects. *Nat Genet* 1994;**7**:463–71.

24 Plaza S, Dozier C, Saule S. Quail Pax-6 (Pax-QNR) encodes a transcription factor able to bind and trans-activate its own promoter. *Cell Growth Diff* 1993; **4**:1041–50.

25 Grindley JC, Davidson DR, Hill RE. The role of *Pax-6* in eye and nasal development. *Development* 1995;**121**:1433–42.

26 Cvekl A, Sax CM, Li X, McDermott JB, Piatgorsky J. Pax-6 and lens-specific transcription of the chicken delta 1-crystallin gene. *Proc Natl Acad Sci USA* 1995;**92**:4681–5.

27 Chalepakis G, Wijnholds J, Giese P, Schachner M, Gruss P. Characterization of Pax-6 and Hoxa-1 binding to the promoter region of the neural cell adhesion molecule L1. *DNA Cell Biol* 1994;**13**:891–900.

28 Mirzayans F, Pearce WG, MacDonald IM, Walter MA. Mutation of the *PAX6* gene in patients with autosomal dominant keratitis. *Am J Hum Genet* 1995;**57**: 539–48.

29 Gerard M, Abitbol M, Delezoide AL, Dufier JL, Mallet J, Vekemans M. *PAX* gene expression during human embryonic development, a preliminary report. *C R Acad Sci III* 1995;**318**:57–66.

30 Matsuo T, Osumi-Yamashita N, Noji S, Ohuchi H, Koyama E, Myokai F, *et al*. A mutation in the Pax-6 gene in rat *small eye* is associated with impaired migration of midbrain crest cells. *Nat Genet* 1993;**3**:299–304.

31 Quiring R, Walldorf U, Kloter U, Gehring WJ. Homology of the *eyeless* gene of *Drosophila* to the *Small eye* gene in mice and aniridia in humans. *Science* 1994;**265**:785–9.

32 Halder G, Callaerts P, Gehring WJ. Induction of ectopic eyes by targeted expression of the *eyeless* gene in *Drosophila*. *Science* 1995;**267**:1788–92.

33 Zhang Y, Emmons SW. Specification of sense-organ identity by a *Caenorhabditis elegans* Pax-6 homologue. *Nature* 1995;**376**:55–9.

34 O'Farrell PH. Unanimity waits in the wings. *Nature* 1994;**348**:188–9.

35 Scott MP. Intimations of a creature. *Cell* 1994;**79**:1121–4.

36 Glaser T, Ton CC, Mueller R, Petzl-Erler ML, Oliver C, Nevin NC, *et al*. Absence of *PAX6* mutations in Gillespie syndrome (partial aniridia, cerebellar ataxia and mental retardation). *Genomics* 1994;**19**:145–8.

37 Mason IJ. The ins and outs of fibroblast growth factors. *Cell* 1994;**78**:547–52.

38 Slack JMW. Inducing factors in *Xenopus* early embryos. *Curr Biol* 1994;**4**: 116–26.

39 Cohn MJ, Izpisúa-Belmonte J-C, Abud H, Heath JK, Tickle C. Fibroblast growth factors induce additional limb development from the flank of chick embryos. *Cell* 1995;**80**:739–46.

40 Peters KG, Werner S, Chen G, Williams LT. Two FGF receptor genes are differentially expressed in epithelial and mesenchymal tissues during limb formation and organogenesis in the mouse. *Development* 1992;**114**:233–43.

41 Peters K, Ornitz D, Werner S, Williams L. Unique expression pattern of the FGF receptor 3 gene during mouse organogenesis. *Dev Biol* 1993;**155**:423–30.

42 Tavormina PL, Shiang R, Thompson LM, Zhu YZ, Wilkin DJ, Lachman RS, *et al*. Thanatophoric dysplasia (types I and II) caused by distinct mutations in fibroblast growth factor receptor 3. *Nat Genet* 1995;**9**:321–8.

43 Jabs EW, Li X, Scott AF, Meyers G, Chen W, Eccles M, *et al*. Jackson–Weiss and Crouzon syndromes are allelic with mutations in fibroblast growth factor receptor 2. *Nat Genet* 1994;**8**:275–9.

16: Molecular genetics of common diseases

Ian N M Day

Common disease

The lay opinion that "it runs in the family" has been subjected to scientific examination from many angles for many diseases. For most common diseases, there appears to be a genetic component operating in conjunction with an environmental component. Currently, it is the turn of the molecular geneticists to unravel the specific genes and human genetic variations underpinning the deductions made from classic genetic studies of families, twins, siblings, adoptees, and population groups. Thus the "black box" approach can now be improved with a study of the "nuts and bolts". Common diseases and traits which represent major causes of mortality and/or morbidity in Britain include coronary disease, hypertension, diabetes mellitus, Alzheimer's dementia, asthma, arthritides, cancers of lung, breast, colon, and prostate, psychoses, and susceptibility to infections. For every one of these, some progress has now been made in identifying genetic factors important in the pathogenic process of the disease, but, for all, the complete picture of the polygenic nature of the disease is still far from complete. Nevertheless, within the next few years we can reasonably expect to see these pictures emerge, and for attention to shift from research into the pathology, to research into new therapeutic strategies. This chapter considers the extent of human genetic diversity, how we can study it, what the nuts and bolts of the human genome are, what is now known about the genetics of some common diseases, and how in the future this knowledge could lead to improved health.

Human genetic diversity

Both classic genetic studies and, more recently, molecular genetic studies show that humans are genetically different. In some instances, these differences are attributable to rare variations ("private" to individuals or families), and in other instances the differences reflect common polymorphism. Not surprisingly, rare but major single gene diseases such as muscular dystrophy and cystic fibrosis represent the former category of genetic variation, whereas common disease susceptibility will frequently turn out to reflect polymorphic variation, because this is the only category of variation that pertains to substantial subsets of the whole population. It may help the reader to have an overview of the nature and evolution of human genetic diversity, to enable a better understanding of the task at hand in unravelling the basis of polygenic diseases. The human genome contains an estimated 3000 000 000 base pairs and an estimated 100 000 genes. Only a small fraction of the genome encodes proteins, but other parts of it are concerned with gene regulation, chromosome architecture, and other functions. Large parts of the genome are, however, probably without function, and many highly polymorphic markers such as microsatellites (probably more useful to the scientist than to the organism, see below) are found within these tracts. Additionally, the mitochondrion has a small genome of its own. Common polymorphism (defined as a variation found at a frequency of at least 1% in the population) is estimated to occur at about one position in every 200 base pairs to one position in every 1000 base pairs. There is less variation in protein coding regions than in non-coding regions. Thus there might be 3 million common variations in the genome deserving of analysis for their relationship with common disease, of which a subset of perhaps tens of thousands gives rise to variations of either protein sequence or level of expression and is therefore of obvious interest for possible functional diversity.

So far only a few thousand genes have been fully characterised, and probably only a few hundred common variations have been examined for their relationship to human traits, disease risk, disease progression, or response to therapy. To be prevalent in a population, a genetic variation must either have had a long time, in terms of generations, to become prevalent, or it must have been subject to strong positive selection bias; the opposite is true for rare variations. It may be, therefore, that much common variation predated modern human speciation, that is, several hundred thousand years, and that negative selection has been insignificant because only complex combinations of common genotypes predispose the disease, the environment was different in times past, or the disease only occurs beyond reproductive age. Our thinking on how combinations of genes or genes and/or environment influence the occurrence of common diseases and diseases of onset in later life is still at a very early stage, but an important point is that this category of research steps away from the study of rare patients and will demand population scale epidemiological approaches, and extensive multi-centre cooperation to achieve sufficient statistical power.

Approaches to research

Family studies

For major single gene diseases, there is an obvious pattern of inheritance within an individual family. For example, for an autosomal dominant disorder, every second member will be affected. First degree relatives (for example, parent–child) have 50% of their genome in common, 50% different (for example, half the child's genome is inherited from the *other* parent). Thus, given enough family members, and a means of examining which part of the genome is common to all affected members but not present in unaffected members, it should be possible to identify the region containing the affected gene. Such studies have now been made routine by the development of microsatellite based "genome scans" (see below). Such studies may highlight a gene important to a particular function, which may also contain common variations of lesser impact in the general population, but there is no prior reason to be certain that this will turn out to be the case.

Studies of affected sib pairs

Identification of disease genes by family studies is made simplest where there is complete penetrance of the gene effect, where the number of family members is sufficiently great, or where data from several families can be pooled because the disease in each family is linked to the *same* locus (that is, the disease shows no locus heterogeneity). In many diseases, such as diabetes mellitus type I, there is a significant genetic component, but the penetrance of effect of a given gene is modulated by other genes and environment, and different genes may be involved in different cases. This precludes family linkage studies, but an alternative is to collect large numbers (hundreds) of affected sib pairs for a disease. For each sib of an affected sib pair, the reason that they are both affected is that they share some disease predisposing gene(s) inherited in common ("identical by descent") from their parents. Within the collection of pairs, the entire genome can be scanned for any regions which are identical in both sibs more frequently than expected by random chance. As for family studies, the "scan" analyses intergenic polymorphic microsatellite markers rather than genes themselves. The sib pair strategy avoids the problem of incomplete penetrance and inability to get sufficient study size with individual families by selecting many unrelated but *affected* sib pairs, and the statistical tests for regions of the genome identical at greater than chance frequency are tolerant of locus heterogeneity. Recent studies in type I diabetes mellitus have implicated several genomic regions in its pathogenesis. Nevertheless, the resolution of the approach for determining cumulative small percentage effects of large numbers of genes (that is, polygenic rather than oligogenic effects) may still be limited by the number of affected sib pairs who can be found for study. The study of more remotely related affected relatives, sibs discordant for disease, and the determination of the parental genotype *not* inherited as a control, represent other related study designs.

Population association studies

Functional variations within genes, for example, some variations resulting in a different sequence of a protein, can be tested directly in unrelated individuals. Thus, rather than looking at functionless microsatellite polymorphic markers to determine allele sharing in affected relatives, effect on diseases or traits can be measured directly in unrelated individuals. Thus case control comparisons, or comparison of occurrence of a disease or trait (qualitative or quantitative) for specific genotypes, can be undertaken. This approach was successful in the confirmation of the important role played by variation in apolipoprotein E in the development of sporadic Alzheimer's dementia (see below).

Linkage disequilibrium

Consider two sites in the genome, for example, 1000 bases apart. Suppose that at both sites there is common genetic diversity, desigated A/a at one site, B/b at the other. When variation first occurred at the B site in human evolution, then the new variant b

must have occurred on a chromosome bearing either A or a. Suppose it was on a chromosome bearing a, then initially, in the next generation, wherever we find b, we are likely to find the a genotype at the nearby site on the same chromosome. The markers a and b are said to be in linkage disequilibrium or allelic association with each other. With further generations, meiotic recombinations between the two sites will eventually randomise any allelic association of a with b, as crossovers with chromosomes bearing B accumulate. However, recombination between sites 1000 bases apart would only be expected, on average, once in every 100 000 meioses. Thus it should be evident that extensive disequilibrium will be found in the human genome. This is most evident for closely spaced polymorphic markers and for newer mutations or variations for which there has been insufficient opportunity for meiotic recombinations. Linkage disequilibrium has been extensively used to map rare disease genes in founder populations, for example, in Finns, and has potential for application to the genetics of common disease also.

Types of genetic variation

Human genetic variation takes a number of different forms at the molecular level, and each form has different implications both for analysis and for disease effects.

Base substitutions, small insertions, and deletions

Variations of a single nucleotide (A, C, G, or T) to a different one are common, occurring on average at least one position per kilobase. Within protein coding regions, such a change may alter a codon so that a different amino acid is encoded, or so that a premature stop codon occurs in the protein sequence. These variations seem to reflect a combination of DNA polymerase synthesis errors of base incorporation and failure of mismatch repair machinery. Where an amino acid is changed to another amino acid, the change may have little consequence for the protein, it may disable its function, a "haploinsufficiency" effect (for example, some low density lipoprotein gene mutations in familial hypercholesterolaemia), it may further activate the protein (for example, in the G protein coupled parathyroid hormone receptor in metaphyseal chondrodysplasia), it may vary the performance of a protein (for example, apoE2, E3, and E4 phenotype influence on plasma cholesterol clearance), or it may cause the protein to exert a "dominant negative" effect (for example, the structural disorder created in the myocyte by missense mutations in genes causing dominant hypertrophic cardiomyopathies). The examples cited represent mainly rare single gene diseases, but the classes of effect augur the patterns we may expect for common variations which will turn out to influence the occurrence and course of polygenic diseases.

The dinucleotide sequence 5'-CpG (that is, a G following a C) undergoes a natural chemical modification at many CpG sites in the genome; this natural chemical modification makes it much more prone to mutation, usually to TpG or CpA. In fact about 20% of all base substitutions occur at CpG sites, which is about a twentyfold overrepresentation over what would be expected. This is noted among the spectrum of mutations for many rare diseases, and we would also expect that these hotspots will form an important source of the genetic variation underpinning polygenic diseases.

Single base variations are commonly typed using either a restriction enzyme (for example, the enzyme *Eco*RI could cleave the sequence GAATTC but not any single base variation of this), the fragment sizes being checked by electrophoresis, or using short chemically synthesised complementary oligonucleotides which will only remain bound if they have a perfect sequence match. One objective now is to improve on these techniques in order that millions of such sites in thousands of patients can be studied as candidates in polygenic diseases, hopefully before the turn of the millennium.

Sequence variations involving a few bases are quite common, especially outside coding sequence, for example, in gene control regions, introns, and intergenic regions. Lower complexity sequence may be more prone to such events, and apparently reflects nascent strand slipped mispairing with its template. Within coding sequence the effect (as for premature stop codons) will commonly be more profound. Frequently no useful protein can be synthesised. Some proteins are more important than others, and for some proteins we need the product level achieved by both chromosomes, whereas for others we can get by with half normal level. For example, complete analbuminaemia has few clinical consequences, whereas haploinsufficiency (half normal level) at the low density lipoprotein receptor locus leads to twice normal cholesterol levels, and frequently atherosclerosis, angina, myocardial infarction, or death between age 30 and 60 years. The last thus manifests as an autosomal dominant condition. By contrast, half normal level of phenylalanine hydroxylase activity does not cause clinical disease, but individuals who inherit two defective gene copies will lack any phenylalanine hydroxylase activity and will display phenylketonuria. This is an example of a classic autosomal recessive pattern of inheritance. The concepts of dominance and recessivity (and also of X linkage) should translate into quantitative and threshold effects in polygenic disease, with the caveat that interaction with other genes and environment may be in an additive, multiplicative, exponential, or other model according to the mechanism of the underlying biological interactions.

Minisatellite and microsatellite repeat sequences

At many sites in the genome, simple repeat sequences are found, for example, repeat sequences of the form $(CA)_n$. Such dinucleotide (and tetranucleotide) repeats often display 10 or more different length alleles, that is, each locus is highly polymorphic in the population. The reason for this is assumed to be that there is a slight tendency (perhaps once every few thousand generations) to slipped strand mispairing during meiosis and that therefore substantial polymorphism has accumulated in human history. Although these markers are not themselves genes, they enable the ready tracking of regions of the genome in family members, and the testing for chromosome regions that are identical by descent in affected sib pairs. For these reasons, complete genome maps have now been built up for these loci, enabling "genome scans" of each and every region of every chromosome for family and sib pair studies as described above. Such genome scans pinpoint the region of the genome in which to search for a particular disease gene locus.

Minisatellites consist of repeats of longer sequence motifs, for example, 10–30 base pairs, and these loci tend to be extremely polymorphic, although less convenient to analyse than short repeat sequences. First identified by Alec Jeffreys, their utility was immediately recognised for specific identification of individuals (for example, forensics, questions of parentage). It is possible that there will be further use in the study of populations and diseases.

Triplet repeat sequences

Several rare gene disorders such as Huntington's disease and fragile X syndrome have turned out to be caused by length expansion of triplet repeats, which results in the disruption of gene function. Such expansions are "dynamic", in that the length expansion can range from normal, through permutation or carrier states with little or no clinical effect but predisposing further expansion in the next generation, to large highly deleterious expansions. Clinical effects may partly correlate with the size of the expansion. So far, such effects have only been described in rare major monogenic diseases, but similar repeats are probably common throughout the genome and it remains to be seen whether such variations will also be implicated in polygenic diseases. This category of variation is conceptually new: first in that it provides a plausible molecular basis for the often doubted phenomenon of anticipation (where a disease appears to get worse as it is passed down the generations); and second that it represents a different way in which the genotype–phenotype relationship can be one of a continuous rather than a discrete variable.

Major structural rearrangements

Duplication is a common theme in biology, and the genome is no exception. Many regions, such as the HLA region on chromosome 6, contain loci representing duplications of an ancestral gene. As evolutionary time goes on, the sequences and roles of the duplicated loci may diverge. Such regions do, however, remain prone to rearrangements by one of several possible mechanisms, but each involving one locus interfering with the other. These rearrangements may ablate genes, or join two genes into one. Such events display a remarkable range of clinical effects. At the α globin loci, two gene copies on each chromosome is normal, totals of three, two, one, or no normal copies respectively result in thalassaemia trait, HbH disease, or hydrops fetalis. Where malaria is endemic, balancing selection favouring survival with the malaria of thalassaemia carriers ensures high frequency of the deletion variants. Copy number and hence expression level of the closely spaced complement 4A and 4B genes on chromosome 6 influence overall complement activity, and this in turn seems to influence survival after myocardial infarction. There is considerable potential for both quantitative effects and qualitative effects in this category of genetic diversity, and it is a reminder that even whole genes can be "relatively" dispensible, particularly where they belong to duplicated gene families.

Sex dependent effects (X chromosome, mitochondrial DNA, and imprinting)

Males have only one X chromosome, whereas females have two X chromosomes. Thus gene deficiencies involving X chromosome genes, such as Duchenne muscular dystrophy (DMD), usually present as disease in males only. One would expect that the effects of about 5% of genes will be sex dependent on account of their genomic location, although the effects of many others will be modulated by variations in physiology implicit in difference of sex. DMD exemplifies the first, and BRCA1, a familial autosomal dominant breast cancer gene, is a good example of the second.

Diseases associated with mitochondrial genes such as mitochondrial myopathies illustrate a different pattern. Mitochondria, with their own independent genome, are transmitted only from the mother via the oocyte, whereas the sperm makes no mitochodrial contribution at fertilisation. As a consequence, functions encoded by the mitochondrial genome exhibit maternal inheritance.

Copies of at least a few parental genes are selectively expressed by one parental allele and not by the other, termed "imprinting". For example, insulin like growth factor II is expressed by the paternal but not by the maternal allele in the fetus. Teleologically, this has been considered as a mechanism to protect the mother against fetal overgrowth which would have adverse consequence for mother, fetus, or other offspring.

Thus, for several different reasons, we can expect that many genes influencing the course of common disease will have to be examined separately in males and females.

Some common diseases: current knowledge

Three common diseases, type I diabetes mellitus, Alzheimer's disease, and coronary disease, illustrate well both the different approaches described above and different mechanisms of polygenic disease. Owing to constraints on space, current knowledge for some other diseases is presented in summary form (table 16.1).

TABLE 16.1—*Genes implicated in some common diseases*

Common diseases	Genes implicated	Comments
Alzheimer's disease	*ApoE* gene	Results replicated in many different studies
Type I diabetes	HLA, insulin gene, loci on 15q 11q13, 6q25, 18q, 2q31, 6q27, 3q21–25, 10q11, 2–q11.2 (loci IDDM3–10)	Genes at loci 3–10 not yet identified
Coronary heart disease	Apo(a) gene, lipoprotein genes, clotting factor genes	
Asthma	IgE receptor (FcERI-β) Leu181/ Leu183, interleukin 4 cluster, TCR-α	Maternal inheritance effect. Would not account for more than 5% of asthma
Osteoporosis	Vitamin D receptor	
Malaria resistance	Sickle cell (β globin) gene, thalassaemia (α globin) genes, glucose-6-phosphate dehydrogenase B and C protein; blood group, Duffy and O antigens; tumour necrosis factor α; HLA class I and II diabetes	
Autoimmunity syndromes	Major histocompatibility complex (MHC)	

Type I diabetes mellitus

Type I, also called insulin dependent (IDDM) or early onset, diabetes mellitus, affects about 0·2% of children. Given a child with IDDM, the risk that a sibling will develop IDDM is about 15 times greater than the general population risk. This points to the influence of genetic factors. However, only one third of identical twins concord for IDDM, pointing also to the importance of environmental factors. Many years of biochemical and immunological research have led to the recognition that IDDM is an autoimmune process involving destruction of insulin secreting pancreatic β cells. Features that are presymptomatic include activated T lymphocytes expressing HLA-DR antigen, islet cell autoantibodies, and insulin autoantibodies. It is therefore no surprise that genetic diversity in the major histocompatibility complex and insulin gene loci shows association with diabetic risk, and these loci have been closely examined as candidates using a variety of approaches.

Very detailed analyses in the major histocompatibility complex, including comparison of HLA genes at the sequence level, led to the identification of an important codon at position 57 in the HLA-DQ β chain, likely to be involved in antigen presentation in anti-islet response. Polymorphism in the β chain of HLA-DR, apparently at codon 74, also represents a risk variable for IDDM. For example, the susceptibility allele DQB1*0302 confers no risk in the presence of protective allele DRB1*0403. Certainly, insulin B chain is an important autoantigen, whose binding to HLA-DR is influenced by DR β chain codon 74; thus the protective allele is very probably itself functional rather than a linkage disequilibrium marker for some other functional variation. However, the large number of genes and variations in the MHC region makes it very complex to distinguish truly functional variations from disequilibrium markers.

The insulin gene has also been examined in great detail. Association studies of polymorphism in the insulin gene gave the earliest positive results, but these were not confirmed initially in family studies. There was therefore concern that the apparent association could be a result of population stratification. This can occur if a population is incorrectly assumed to be homogeneous, when it actually contains different subpopulations with different allele frequencies. Then, if the disease has its origins in some genotype feature of one subpopulation, any genotype marker of that subpopulation, on any chromosome, will tend to show positive association with the disease, whereas only one particular genotype at one locus is truly functional rather than simply being a confounding factor. The haplotype relative risk method, which types the non-transmitted parental genotypes as "unaffected artificial controls", is in effect a family based association test which avoids the population stratification problem (because the genotype comparisons are of cases and "controls" obligatorily from the same subpopulation). This method was first used to prove conclusively that the association of the insulin gene region with type I diabetes represented linkage disequilibrium not population stratification. Study of polymorphisms, including single base diallelic polymorphisms spanning the insulin gene and flanking regions, out to the nearby flanking tyrosine hydroxylase and insulin like growth factor II genes, and also a variable number tandem repeat (VNTR) in the promoter region of the insulin gene, defines a "peak of risk" around the insulin gene promoter region. The VNTR does seem to affect insulin gene expression, but no causal chain between it and development of IDDM has yet been put forward.

Although these candidate gene studies have continued, it has become possible to undertake complete genome scans using VNTRs to identify genomic segments identical by descent in affected sib pairs, and to test statistically for segments identical by descent more frequently than chance predicts, within a collection of sib pairs. This approach has been exhaustively applied to a collection in excess of 200 IDDM sib pairs, resulting in the identification of at least 10 genomic regions which appear to confer an increased relative risk of IDDM. These regions have been designated IDDM1, IDDM2, IDDM3, etc. Reassuringly, IDDM1 is the MHC region on chromosome 6p21 and confers the strongest risk (2·6-fold or, calculated another way, explaining 35% of the familial clustering of IDDM). Also, IDDM2 is the insulin gene region on chromosome 11p15 with a relative risk of 1·3, accounting for 10% of the familial clustering observed. Other loci seem to be located on chromosomes 15q, 11q13, 6q25, 18q, 2q31, 6q27, 3q21–q25, and 10p11.2–q11.2, termed respectively IDDM3 through IDDM10. Each locus will make a small genetic contribution not dissimilar from that of variation at the insulin locus. Refining these map positions beyond about 10 centimorgans (cM) (perhaps 10 million base pairs containing up to 1000 genes) demands more sib pairs (thousands) than can be found or collected. In the absence of obvious candidate genes or even gene maps in these regions, it remains to be seen how much refinement may first be achieved using linkage disequilibrium (allelic association) mapping (see above). Linkage disequilibrium would not usually extend much beyond 1–2 cM, being minimal in old populations such as Africans but more evident in "new" populations such as Finns. Thus it should be possible to home in to regions containing about 1 million base pairs. Detailed search by association, for "peaks of apparent risk" would then have to try to take account of the many thousands of diallelic polymorphisms and other polymorphisms in such regions, until a candidate gene emerged either by this method or by candidacy based on gene function deduced from sequence, pattern of expression, or other model.

These studies demonstrate a systematic but laborious approach that is now feasible for dissecting the oligogenic or polygenic basis for a common disease which is phenotypically well defined, occurs relatively early in life, and seems to have a clearcut pathway of pathology.

Alzheimer's disease

First recognised in 1907 by Alois Alzheimer, this disease is now known to be a common disease which is the predominant cause of dementia over the age of 65 years. Progression of mild short term memory problems, through to massive loss of memory, language, and orientation, leads to incapacitation of the individual. With increased survival to old age, society is faced with the future care of vast numbers of dementing patients whose quality of life is poor. Evidence gathered from multiplex families with late onset dementia suggested that familial clustering was unlikely from chance alone and, in 1991, an affected pedigree member analysis was used to show that a region of chromosome 19 was common to affected pedigree members more frequently than random chance would predict. This linkage was confirmed, but a much earlier report of an association of a restriction fragment length polymorphism allele of the *APOCII* gene with Alzheimer's dementia did not replicate in these data. The *APOCII* gene is adjacent to the *APOE* gene on chromosome 19: the failed replication of the *APOCII* gene association, and the fact that apoCII and apoE were regarded as lipoprotein components relevant to atherosclerosis diverted attention from these genes as candidates for the Alzheimer locus. However, apoE was shown to bind Alzheimer amyloid β peptide and, furthermore, apoE antisera stained senile plaques and neurofibrillary tangles, prompting investigation of *APOE* genotype in Alzheimer's disease. There are three common isoproteins of apoE, designated 2, 3, and 4, with allele frequencies of 6%, 78%, and 16%. The differences are at amino acids 112 and 158, the isoforms containing respectively, cysteine and cysteine (two), cysteine and arginine (three), or arginine and arginine (four). The association of *APOE* genotype with susceptibility, *APOE4* marking increased risk, has now been confirmed in many laboratories, in family and sporadic late onset disease in many studies in various racial groups. The effect of *APOE4* is dose dependent, that is, E4/E4 genotype has the most severe effect, advancing age of onset by about eight years per allele from an age of 84 years if no E4 allele is present, to 68 years for E4/E4 genotype. *APOE2* seems to mark reduced risk, relative to E3. However, the observed associations do not prove that these genotypes are the functional feature on chromosome 19; they could simply be acting as linkage disequilibrium markers (see above) for other genetic diversity in a nearby gene which causes the true pathological effect. However, the following leave *APOE* genotype with very strong candidacy for representing the functional site within this genomic region of chromosome 19: the circumstantial evidence; the expression of apoE in the nervous system; the strong enhancement of expression in response to nerve injury; the difference in expressed isoproteins (which at least in the cardiovascular system confers differences in interactions, for example, with lipoprotein receptors); and differential effects of isoform specific, β, very low density lipoprotein on neurite extension in vitro. Functional studies or an animal model may prove the assumed pathogenic role of apoE4, which may be an important determinant for 50% of cases of late onset Alzheimer's dementia. ApoE genotype seems also to influence the age of onset of dementia in the rare families developing Alzheimer's dementia at a very early age, who have an autosomal dominant inheritance caused by mutations in the *APP* gene on chromosome 21. ApoE genotype also seems to interact with head trauma in the development of Alzheimer's disease, and there is also the suggestion of interaction with genotype of the very low density lipoprotein receptor locus. It has been estimated that APOE4 genotype predicts a sixfold greater risk of development of late onset Alzheimer's dementia, but many young adults with this genotype will not go on to develop dementia. Ultimately, refined risk prediction and understanding of pathogenesis may hinge on an array of environmental factors and genotypic variations acting in combination.

The history of our current understanding of the molecular genetics of late onset Alzheimer's dementia provides reassuring evidence that the genetics of a complex disease can be unravelled, leading to the implication of a gene not obvious at the outset of the research. It also illustrates the complex interplay of family, association, and functional studies in such research.

Coronary heart disease

Coronary heart disease is a final common pathway of disease which can involve, in various combinations, smoking, hyperlipidaemias and atherosclerosis, hypertension, diabetes, variations in clotting, oxidation status, and perhaps resistance to fatal arrhythmias. Compared with some of the other common diseases considered, it is therefore expected to be highly polygenic and complex, ranging from relatively rare single gene disorders such as familial hypercholesterolaemia, involving LDL receptor gene defects (1/500 prevalence), to cumulative contributions from many genes each having perhaps 1% impact on the genetic risk. On account of extensive research on many measured intermediate traits, particularly easy as many are plasma constituents, many genetic components have been under study for over 10 years, and previous editions of this chapter have summarised much of this work. However, there is the prospect that the full polygenic picture will only emerge when every variation in the entire human genome has been examined; possibly, even then, the combinatory nature of gene interactions (epistasis) may preclude any study size from determining the effects of every combination! Thus we can anticipate that, although research into the polygenic components of coronary disease had an early start, it will be a late finisher in view of its great complexity.

How are we to establish whether a particular gene contributes to the genetic variation in a trait such as blood pressure or plasma cholesterol concentrations? Genetic epidemiology has provided statistical methods for measuring the effect of genetic variation on a phenotypic trait in a population (the so called measured genotype approach). For example, functional polymorphism of apolipoprotein E (encoded on chromosome 19q13) can be assessed by isoelectric focusing or by the analysis of *apoE* gene polymerase chain reaction (PCR) products. In the normal population these three polymorphisms have a substantial effect on the normal variation in plasma lipid concentrations; they account for 16% of genetic variation in cholesterol concentrations in the population. These small changes in plasma cholesterol concentration are sufficient to produce a substantial enrichment of the E2 allele with advancing age and a decrease of the E4 allele in the ageing population, perhaps as a result of an increased number of deaths from myocardial infarction (see section on Alzheimer's disease). Very large case control studies show the expected genotype bias with respect to coronary events. Yet the effect of these polymorphisms on individual lipid concentrations and risk of coronary heart disease is small.

The exception to this is when homozygosity of the E2 allele occurs in association with diabetes mellitus, hypothyroidism, or another genetic defect of lipid metabolism. This leads to the condition of dysbetalipoproteinaemia, which is characterised by profound hypertriglyceridaemia, moderate hypercholesterolaemia, and premature coronary artery disease. This condition is caused by defective clearance of chylomicron remnants (caused by the E2 allele, which does not bind to the low density lipoprotein or chylomicron remnant receptor) combined with another condition that produces hyperlipidaemia.

A remarkable size polymorphism is shown by the apolipoprotein (a) gene (apo(a)). Apolipoprotein (a) resembles the precursor of the enzyme, plasmin, which is responsible for the hydrolysis of fibrin clots. Apolipoprotein (a) is secreted tightly associated with apolipoprotein B100 from the liver, where two molecules together form a lipoprotein particle designated Lp(a) lipoprotein. The apo(a) gene has a variable number of repeats of the region coding for kringle (a disulphide bonded domain shaped like a Danish cake) IV. This gives rise to extreme size variation from 280 to 830 kDa. The size is inversely correlated with Lp(a) lipoprotein levels in the circulation, and individuals with the small protein have much higher concentrations and are at much greater risk of coronary heart disease. The apo(a) gene size polymorphism accounts for more than 90% of the variation of Lp(a) lipoprotein levels in the population. Although the cause of the increased likelihood of disease is not certain, Lp(a) lipoprotein probably in some way interferes with normal endothelial cell fibrinolysis and promotes a procoagulant state. Raised blood Lp(a) lipoprotein concentrations with raised cholesterol concentrations synergise in increasing the risk of coronary heart disease.

Future management possibilities

Identification of targets for drug development

Pharmaceutical companies, as well as large academic genome laboratories, have taken much interest in developing complete catalogues of cDNA clones derived from messenger RNA, because these catalogues will represent each and every gene, in a form from which the protein sequence can be deduced and perhaps expressed. Functionally expressed proteins open many opportunities for the screening and testing of traditional pharmacological agents. This strategy focuses on possible drug targets. However, genetic diversity underpinning common disease may point to the most appropriate targets for drug development. For example, we now know that expression of apolipoprotein E4 predisposes Alzheimer's dementia in a dose dependent fashion, and also that, at least in vitro, apoE interacts with the Aβ hydrophobic peptide component of neurofibrillary tangles in an isoform dependent fashion. A drug modulating this interaction, perhaps in an isoform selective way, would be one possible approach to prevent or reduce development of Alzheimer's dementia. On the same theme of structural interactions, vaccine development may be enhanced by a more detailed understanding of the important genetic variables in the HLA gene classes important in antigen recognition and presentation.

Diagnosis and prognosis

In the preceding section, an imaginary possibility for the future prevention of Alzheimer's dementia was presented. If we had such a therapy, we would then also implement a diagnostic test so that it could be selectively prescribed to those with an apoE4 allele. Diagnosis means the identification of the underlying cause of a patient's clinical condition. Many genetic predispositions are known, such as the influence of the HLA system and insulin gene promoter region in type I diabetes, the interrelationship between HLA type and other autoimmune diseases, the role of apoE in dementia, and apparently the effect of variation in the angiotensinogen gene on occurrence of hypertension. Unfortunately, although these markers help to paint the picture of the underlying pathology, they do not yet influence patient management in any way that makes diagnostic tests worthwhile. However, a better case may be made for the identification of glucokinase gene variants in diabetics. Glucokinase represents the rate limiting step in glucose metabolism in the pancreatic β cell. Glucose metabolism determines ATP level, which in turn influences potassium and calcium flux in the cell, and this affects rate of insulin secretion. In glucokinase deficiency, the consequence is resetting of glucose level, to a level consistent with a diagnosis of diabetes, but stable and with mild if any clinical effect. This represents one category of maturity onset diabetes of the young (MODY): ketotic crisis does not occur, these patients do not need insulin, and the outlook is good. Appropriate diagnosis averts mistaken recognition as insulin dependent diabetes, and potential lifelong inappropriate insulin therapy. In addition, the prognosis is good and the patient can be appropriately advised, and the autosomal dominant nature of the disorder allows genetic counselling within the family. At present, the DNA technology is insufficiently effective to attempt to classify all diabetics for glucokinase gene variation, so that only carefully selected groups can be scanned. Several MODY loci exist in the genome, distinct from the set known to predispose to type I diabetes: in the future the genetic components of type II diabetes will be unravelled. Among late onset diabetic individuals, a proportion is controlled by diet, a proportion by oral hypoglycaemic agents, and the rest by insulin. Management of dyslipidaemias is a further consideration. The main long term concern is the risk of end organ damage, particularly to vascular system, kidneys, and eyes, which may not solely reflect glucose levels. With the glucokinase version of MODY in mind, it is not difficult to imagine how future molecular classification could refine management of the individual patient. Diabetes represents a phenotype definition that may cover a range of different pathologies with different natural histories and different therapeutic needs.

Monitoring, screening, avoidance of environmental risk in susceptible individuals, and family tracing of treatable disease or preventable risk

Monitoring of disease applies to somatic cells rather than to germ cells; this is particularly pertinent to cancers, where tumour burden, therapeutic response, or new cellular change can be characterised at the genetic level.

Screening is an attractive possibility for single gene disorders such as cystic fibrosis and acceptable provided that stringent criteria can be met: the natural history of the disorder is known; there is the mechanism and means to improve health; and tools of appropriate sensitivity and specificity are available to undertake such screening. It is less obvious how screening could be applied in polygenic disease, but for asthma, diabetes, or rheumatoid arthritis it might be possible to avert disease development by protecting identified susceptible individuals through "risk windows" in their lives—for example, stringent avoidance of allergens, perhaps insulin to "rest the pancreatic β cells", or aggressive therapy at the outset of disease in selected subsets of arthritic individuals destined for a severe course.

Drug therapy has been shown to be effective both in primary prevention (that is, before coronary disease ever occurred) and in secondary prevention (that is, averting a second coronary event) in hypercholesterolaemia. In some single gene disorders such as familial hypercholesterolaemia, where drug therapy is available, systematic public health programmes testing (genetic), tracing, and treating family members could be justified. Major risk genes might be characterised in families in this way in the future.

Genostratified drug trials and classification by genotype of those most likely to benefit from therapeutic modalities

Hypertrophic cardiomyopathy affects only a small number of individuals. Autosomal dominant forms involve one of several structural genes of the cardiac myofibril, predisposing both hypertrophy and arrhythmias which may be fatal. In families where the mutation has been determined, there is a definite means to identify those at risk, not always possible by other methods. However, apart from the avoidance of severe exercise throughout life, it is unclear how else to reduce risk of sudden death in early adulthood. Several categories of drugs, such as angiotensin converting enzyme inhibitors to reduce hypertrophy, β blockers, and antiarrhythmic agents, would be candidates for prophylactic therapy. Most physicians would consider genetic diagnosis inappropriate in the absence of significant therapeutic benefit. The most logical approach would be controlled drug trials on genotype determinate individuals: the genotype being a research tool only at this stage, it would not be reported to the patients. It can be imagined that similar approaches might be appropriate for groups of patients with cancer risk genes. In the future, common genetic variants such as HLA type, asthma risk genes, etc could be used to stratify or select patients at entry to drug trials. By selecting those at higher risk, and hence most likely to benefit from the trial drug, it would be possible to make trials smaller (fewer patients needed), faster (because selection of more likely responders would more quickly result in endpoints), cheaper, and safer (because fewer people need be exposed to the trial drug).

Given information about genotype dependence of specific drug responses, genotype would then become a prognostic tool for more specific tailoring of drug therapy. For example, it is well known that Afro-Caribbeans have completely different responses to white Europeans in hypertension, but are much more resistant to myocardial infarction. Chinese individuals are commonly susceptible to severe flushing given alcohol, as a result of a genetic deficit of aldehyde dehydrogenase.

It is highly likely that within populations, as well as between populations, there will be genetic variations influencing drug target responsiveness, susceptibility to categories of side effect, drug metabolism, etc. In some instances, trial of therapy is appropriate (for example, where there is a measurable short term response), but for long term and prophylactic management predictive tools would be very helpful.

Cloned gene products

Blood proteins such as clotting factors, hormones, and immune factors are obvious candidates for in vitro expression. The protein does not have to be introduced into cells, but simply into the bloodstream. Additionally, this also circumvents the risks of infectious agents when using natural blood products or purified products, for example hepatitis viruses, human immunodeficiency virus, and Jakob–Creutzfeldt disease. Replacement of major deficiency, for example, in haemophilias, is obviously justifiable, but what is the limit of partial deficiency? Short stature is a good example. Height that is three standard deviations below the mean usually turns out to represent identifiable underlying pathology, but height two standard deviations below the mean may well not lead the paediatrician to any particular diagnosis. One differential diagnosis is deficiency somewhere in the growth hormone axis (growth hormone releasing hormone, its receptor, growth hormone, growth hormone receptor, insulin like growth factor, and its receptor). Pituitary dwarves with growth hormone deficiency have very short stature and respond to growth hormone. It has recently been shown, however, that some "short normal" individuals have mild growth hormone receptor defects with partial function, and these individuals should respond to higher doses of growth hormone than their own pituitaries produce. What constitutes an indication for treatment—height as such, proven genetic variation, or simply purchasing power? Body builders are interested in a genetically modified form of insulin like growth factor; most is inactive in plasma because it is attached to a binding protein, but a genetically engineered form has been expressed that remains unbound and hence active. In the next few years, we may well come to understand the main genetic determinants of height and weight, ranging also through a spectrum of pathological and cosmetic sequelae. It will be a matter of public choice to decide how much therapeutic "normalisation" would be acceptable. The mouse "ob" gene seems to encode a satiety factor, and if deficient leads to very obese mice. Gross obesity or dwarfism would be considered pathological by most, but what is to be said to parents who present their "short normal" son at clinic while leaving their "short normal" daughter at home?

Allele specific intervention

Therapy using oligonucleotides is under intensive investigation. There is the possibility both of making modified oligonucleotides which would enter cells and bind, by base pairing, to a specific messenger RNA, or alternatively of oligonucleotides entering the nucleus and modulating the expression of a specific gene. In either instance, the outcome could be to suppress or enhance the expression of a specific protein. In diseases such as Alzheimer's dementia and various amyloidoses, a major feature can be the accumulation of a specific protein as pathological deposits. Reduction of expression would be an obvious way to eliminate the problem. As oligonucleotide base pairing has the capability of distinguishing targets that differ by a single base pair, there would also be the possibility of switching off (or on) a mutant allele, while leaving the normal allele to continue its usual function.

Gene therapy

"Gene therapy" is an exciting phrase that has led to both publicity and confusion. It implies therapeutic manipulation of the genome, a possibility dismissed by the Nobel Laureate and co-developer of penicillin, Chain, as science fiction as late as 1977. This manipulation is being attempted for various diseases and in every category it is somatic cells that are the target. Germline manipulations are not feasible yet in humans, and are certainly not under consideration at present. Nevertheless, in the future, who could say that a perfect repair of a major gene mutation in the germline would be a worse approach to management than the current approach of selective termination of pregnancy?

A typical strategy of somatic cell therapy is to clone a functional gene into a disabled, non-pathogenic virus capable of entering or incorporating into the selected cells, such that the absent function is restored. This has been attempted to replace LDL receptor function in the liver cells of homozygous deficient patients who have no means of clearing atherogenic, cholesterol rich LDL from their plasma. The natural history of such patients is coronary heart disease and death in childhood and early adulthood; they are generally unresponsive to cholesterol lowering drugs and therefore can only be treated by frequent removal of LDL by physical methods such as plasmapheresis. Similar trials are under way to introduce functional product of the transmembrane conductance regulator gene into the bronchial epithelium of cystic fibrosis patients, who are deficient in this gene product. At present such therapies are directed at rare single gene disorders, but, in principle, there could be extensions into the genetics of common disease (for further discussion, see chapter 20).

Ethical and psychological considerations

The possible future advances for improved health considered above concern the benefit to the individual. However, the prognostic power implicit in genotype is of interest also to the insurance industry, to employers, and to other agencies. Such agencies have different concepts of equity and benefit, leading to new conflicts of interest between the individual and society. In effect, genotype will become a major new factor that could unbalance what is currently a relatively stable set of rules governing society. In addition, although some individuals might regard forewarning as forearming, others may be psychologically stressed by knowledge of future risk and would prefer blissful ignorance. These issues will have to be resolved, for the whole approach afforded by molecular genetics to preventive medicine will hinge on utilising its predictive power.

Acknowledgements

I am supported by a Lister Institute Research Fellowship. Professor James Scott is thanked for permitting the re-use of one page from an earlier edition of this chapter.

Recommended reading

BMA Professional, Scientific, and International Affairs Division. *Our genetic future. The science and ethics of genetic technology.* Oxford: Oxford University Press, 1992.

Bodmer W, McKie R. *The book of man. The quest to discover our genetic heritage.* London: Little, Brown & Co. 1994.

Department of Health. *The genetics of common diseases.* A second report to the NHS Central Research and Development Committee on the new genetics. Department of Health, UK, 1995.

Kidd KK. Associations of disease with genetic markers: deja vu all over again (editorial). *Am J Med Genet* 1993;48:71–3.

Schellenberg GD. Genetic dissection of Alzheimer disease, a heterogeneous disorder. *Proc Natl Acad Sci USA* 1995;92:8552–9.

Todd J. Genetic analysis of type I diabetes using whole genome approaches. *Proc Natl Acad Sci USA* 1995;92:8560–5.

Weatherall D. Clinical research in the year 2000. *J R Coll Physicians Lond* 1995; 29:7–8.

Special issue on complex genetic diseases. *Trends Genet* 1995;11:463–524.

17: Impact of molecular biology on clinical genetics

William Reardon

The essential function of the clinical geneticist, genetic counselling, comprises three core elements.[1] First, there is a diagnostic element, because without a correct clinical diagnosis, verified by appropriate laboratory investigation, any discussion of prognosis and genetic implications of disease is severely curtailed and may be frankly wrong. Second, there is the estimation of risk that entails consideration, not only of the patient, but also of other family members. The third aspect is a supportive role, ensuring that management and preventive measures available to the patient receive appropriate discussion and exploration, and are implemented to achieve the outcome desired by the patient. In latter day clinical practice, molecular approaches frequently contribute to altering each of these elements of the clinical genetics consultation. The remit of this chapter is to underline the contribution that molecular biology now makes to clinical genetics practice. In this regard it is helpful to be reminded of how molecular advances have altered the environment of clinical genetics by referring to the history and future direction of the specialty.

In Britain the formal establishment of clinical genetics dates from 1946, when the first such clinic was established at the Hospital for Sick Children, Great Ormond Street, London. Many of the clinical problems encountered then, such as mental handicap, cleft lip and palate, neural tube defects, and muscular dystrophy, remain important today. Equally, the fears and anxieties that provoked families to seek expert guidance are largely similar. What has changed is the vastly improved armamentarium that the clinician can now bring to bear upon the clinical problems confronting him or her, in terms of both clinical information and laboratory options available for diagnostic and prognostic purposes. In the 1940s and 1950s the only information available to the clinician was based on phenotypic and pedigree observations. Whilst colour blindness has been known to be X linked since 1911, on the basis of observing the segregation of the phenotype in pedigrees, the assignment of an autosomally encoded characteristic, Duffy blood group to chromosome 1, did not take place until 1968. By 1995, in excess of 3700 genes had been mapped to their respective chromosomes[2] and about 500 clinical disorders had been related to mutations at specific gene loci.[3] Data relating to gene localisation and mutation identification are constantly increasing. Under the auspices of the Human Genome Project, it is projected that the human genome will be completely mapped and sequenced by 2005. This will provide chromosomal localisation and nucleotide sequence for all human genes. The challenge for clinical geneticists now and for the immediate future is to endeavour, in the face of constant information about new genes and new laboratory techniques, to ensure that the beneficial effects of these advances are incorporated into clinical service provision, in a manner that will combine optimal clinical principles with carefully considered molecular applications. A secondary, although no less onerous, responsibility is to avoid the inappropriate application of molecular techniques to clinical situations.

As a means of emphasising the benefits to patients that accrue from appropriate application of molecular techniques to clinical genetics practice, several examples will be presented. Each case has been chosen to exemplify a particular point and to highlight the synergy between clinical and laboratory based practice which currently prevails.

Altering the risk of genetic disease in a family: Duchenne muscular dystrophy

In the example in figure 17.1, I_1 and I_2 presented in 1970, after the diagnosis of Duchenne muscular dystrophy (DMD) in their son, II_1, then aged seven years. There were no tests available to predict whether I_2 was a carrier of the gene for this X linked condition. All that was possible was to counsel them that two thirds of mothers of boys with DMD are carriers themselves, so that if she conceived another boy, there was a one third chance that he would also inherit the condition. They chose to have no

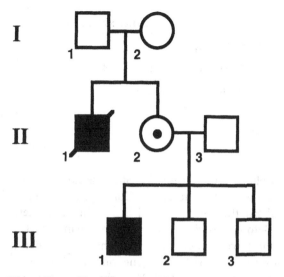

FIG 17.1—*The pedigree, identifying affected males II₁ and III₁, and carrier female II₂.*

further children and their son died in 1977 at the age of 13 years.

The daughter II_2 consulted the genetics clinic in 1981 at the age of 19 years. Empirically, she had a one third chance of being a carrier given that her mother's carrier risk was two thirds. Elevated levels of creatine phosphokinase (CPK) in the blood of carriers of DMD was, by this time, being used as an aid to risk alteration in counselling. Although imperfect in identifying carrier state because of an overlap between the CPK distribution in the DMD carrier population and the normal female population, this investigation was valuable in helping to alter genetic risk in about 60–70% of females at risk of being carriers for DMD.[4] Three separate evaluations of CPK were performed and the average of these indicated that II_2 had a 1:4 (1 in 5) risk of being a gene carrier on enzyme levels. Thus, her risk was lowered as shown in the box.

The effect of the CPK levels was to reduce her carrier risk from two thirds to one third. Consequently she was counselled that if she conceived a son there was a one sixth risk that he would have the condition.

She re-presented in 1987 after the birth of her son, III_1. In view of the concerns arising from the family history, CPK levels were measured in him in the neonatal period. These were very high and a diagnosis of DMD was made. This finding fundamentally altered

217

	Carrier : Not a carrier	
By pedigree	$\frac{2}{3}$	$: \frac{1}{3}$
By enzymes	1	: 4
Final risk	$= \frac{2}{3}$	$: \frac{4}{3}$
	$= 1$	$: 2$
	$= \frac{1}{3}$	

the counselling situation for II_2. She was now a proven obligate carrier of the DMD mutation. Consequently, half of her sons would inherit the condition and half of her daughters would be carriers. Before embarking on another pregnancy, in 1990, she and her husband were counselled that the recent identification of mutations in the dystrophin gene on the short arm of the X chromosome in cases of DMD[5] might be applied as a prenatal predictive test to assess whether another child had inherited the DMD mutation.

Although the exact underlying mutation in dystrophin was still not characterised in this family, an abnormality in the DNA electrophoresis of dystrophin was found in III_1 (fig 17.2). This abnormality helped to differentiate the mutant and non-mutant X chromosomes and to "track" the inheritance of the maternal X in future pregnancies. Thus, in III_2, it was possible to demonstrate, on fetal material, that the pregnancy was that of a male and that he had inherited the opposite maternal X chromosome to his brother with DMD. Hence it was predicted that this boy was 95% likely to be unaffected with DMD, as proved to be the case.

This technique of "tracking" mutations through different affected members in families can lead to errors arising from recombination between the disease gene and the DNA markers being assessed. Accordingly, this risk of recombination is incorporated into the risk counselled to families and accounts for the 5% error rate counselled.

By the time of the next pregnancy, in 1994, the mutation in the dystrophin gene causing DMD in the family had been identified in III_1. Accordingly, the predictive test performed in this pregnancy was to ask specifically whether the fetus had the same mutation or not. The demonstration that the male fetus did not have the mutation causing DMD in III_1 was absolute confirmation that this boy was also unaffected (fig 17.3).

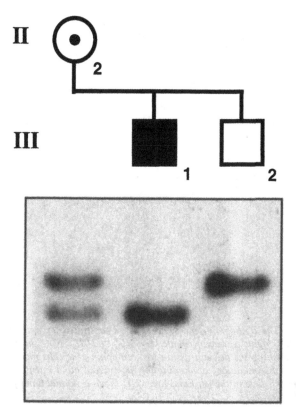

FIG 17.2—*The two X chromosomes in II₂ are identifiable. The lower band is associated with the mutant X and has been transmitted to III₁. As III₂ has inherited the opposite X chromosome, he is unlikely to be affected (see text).*

Commentary

The example presented reflects changes in practice of risk evaluation and predictive testing for Duchenne muscular dystrophy over several years, and reflects the gradual refinement of the application of these advances in the clinical situation. It is still not possible to offer mutation analysis as a predictive test for all DMD families, because of the technical challenges in analysing the gene in any family. Each family has its own mutation and, although there are "hot spots" for mutation within the gene, the identification of the mutation prevailing in the family is crucial before offering a 100% accurate predictive test.

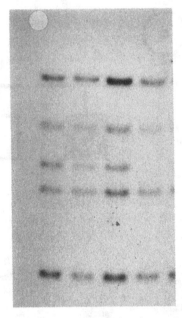

FIG 17.3—*The dystrophin gene analysis is demonstrated for four individuals. From left to right, a normal female, II_2, III_3, and III_1. Note the absent band in III_1, denoting the deletion within his dystrophin gene. His mother, II_2, is a carrier for this mutation, accounting for the reduced band intensity relative to the normal female in the left hand lane. III_3 shows a normal band pattern and is not deleted.*

Confirmation or exclusion of genetic disease in situations of clinical uncertainty: myotonic dystrophy

Myotonic dystrophy is an autosomal dominant condition. In addition, it is one of the few diseases that show "anticipation"— it gets worse from one generation to the next. This phenomenon accounts for the widely varied clinical picture presented in some families. The family demonstrated is typical.

In the example in figure 17.4, both parents in generation I were alive and well in their sixties, when the family came to the attention of the clinical genetics service in 1988. Patients II_1 and II_3 were diagnosed in early adult life with classic signs of weakness: ptosis, myotonia, and early onset of cataract. Although the diagnosis was not in doubt, it was unclear from which parent the gene was inherited. Neither I_1 nor I_2 had any evidence of muscle weakness or myotonia. EMG examination in both was normal and showed

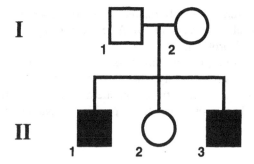

FIG 17.4—*Family with myotonic dystrophy, showing the clinically affected individuals II₁ and II₃. One or other parent must carry the mutant gene, but the clinical signs are too subtle to be certain which parent bears the mutation for this autosomal dominant condition.*

no evidence of subclinical myotonia. Although neither of them had a history of cataract, slitlamp examination was undertaken to ascertain whether the polychromatic lens opacities often seen in myotonic dystrophy were present. I_1 was found to have these non-specific changes on slitlamp examination whereas his wife had a normal examination. By inference it was thought that he might be the subclinical gene carrier, but this was uncertain.

The importance of ascertaining the parental origin of the mutation in this family arose from the concerns of II_2. At the age of 28 years her clinical and EMG examination were normal. By this age, about 60% of patients inheriting the gene would be expected to be clinically symptomatic.[6] The normal clinical examination thus lowered her risk of having myotonic dystrophy to about 20%. However, slitlamp examination showed bilateral subcapsular white lens opacities, too numerous to be passed as a clinically normal variant. Although by no means diagnostic of myotonic dystrophy, the presence of these ophthalmic findings led to reassessment of the risk that she might have myotonic dystrophy and, if so, be at 50% risk of transmitting the mutation to her offspring. She deferred pregnancy, pending developments in the understanding of the molecular genetics of the condition. Happily the gene was cloned in 1992,[7-9] the findings offering not only a diagnostic test but also an explanation for the intergenerational variability in the clinical presentation of the disease. Essentially, there is an instability within the DNA sequence at one end of the gene such that the quantity of DNA tends to increase from one generation to the next at this locus. It is this increasing number of nucleotides that is usually related to the more severe phenotypes, whereas smaller degrees of DNA expansion are associated with fewer clinical problems.

For our patient, II_2, the direct consequence of this discovery was that it was possible to analyse her DNA for the myotonic dystrophy mutation and to demonstrate that she does not have the mutation (fig 17.5). The basis of her lens opacities remains uncertain.

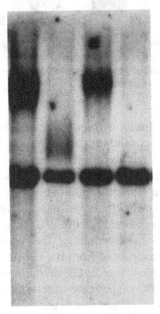

FIG 17.5—*Molecular analysis of the myotonic dystrophy gene in the family shows the expanded mutation in II_1 (lane 1), II_3 (lane 3), and no evidence of expansion in II_2 (lane 4). Note the subclinically affected father (lane 2) has evidence of mutation, although the size of the expansion is less in his case than in his clinically identifiable sons.*

Commentary

This pedigree represents an excellent example of how clinical practice has been altered by advances at a molecular level in the case of some individual diseases. It is no longer necessary to put patients at risk of myotonic dystrophy through the uncomfortable and invasive EMG to arrive at the diagnosis. Although slitlamp examination remains important in myotonic dystrophy, its value nowadays is in determining management of eye complications, rather than in assisting the diagnosis, or, as happened in the case described, confusing it.

Diagnostic cytogenetics in congenital malformation syndromes

About 2% of all children born in the UK have a congenital malformation. Although many such malformations—preauricular pits are a case in point (fig 17.6)—need little or no medical attention

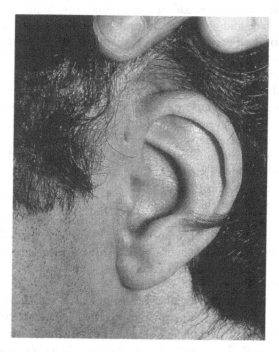

FIG 17.6—*A preauricular pit: generally asymptomatic of itself, in a small proportion of patients this may be the clinical sign betraying an underlying major malformation.*

and do not cause any long term sequelae, sometimes a congenital malformation is a harbinger of other clinical anomalies and may be the prompt to appropriate referral and investigation. A small number of children with congenital malformations have multiple anomalies and may have a recognisable syndrome—the phenomenon whereby a constellation of clinical signs is recognised as occurring together more often than could be explained by chance. Most clinically recognised syndromes date from a single case report, often inappropriately titled and buried in the mists of time by several different authors, often representing different specialties, because the condition has not yet been adequately delineated. Inevitably, clinically diagnosed syndromes are frequently the source of misdiagnosis, individual clinicians disagreeing over whether the dysmorphic features in one child are comparable with those in another. Not only is it distressing to parents that their child's condition is the subject of dispute, but it complicates the counselling situation with regard to future offspring if the diagnosis is unclear. Williams' syndrome represents a case in point. Although the elfin face and coincident murmur were first reported in 1951, it was not until over 10 years later that authors recognised that the children variously described in five different reports with cardiac murmurs, peripheral pulmonary stenosis, idiopathic transient hypercalcaemia, mental handicap, and "characteristic" facial features all represented the same condition, eponymously called Williams' syndrome. Although the phenotypic features were well established, it was not uncommon for a child to be the subject of dispute between clinicians as to whether he or she had Williams' syndrome or not. Essentially these disputes arose because of the absence of a diagnostic test.

That situation has changed since the demonstration that Williams' syndrome is the result of a deletion of the elastin gene on the long arm of chromosome 7.[10] Now it is possible to test using fluorescent in situ hybridisation (FISH) techniques to confirm that this gene is deleted in clinically suspected cases of Williams' syndrome (figs 17.7 and 17.8).

The same technology has been applied successfully to other clinically distressing situations in which the patients have a less apparent dysmorphic phenotype. Consider the family tree outlined in figure 17.9. The family presented to the clinical genetics service for investigation of the developmental delay in their four year old son, II_3. His birth history was normal but early motor milestones were delayed and he also had very significant speech and language delay. He was a non-dysmorphic boy and many of the standard investigations, including routine chromosomes, proved normal. The most striking thing about him was his behaviour, alternating between affection and extreme aggression without apparent provocation. His parents reported that he is highly disruptive in the home. In addition, there was a history of self harming—

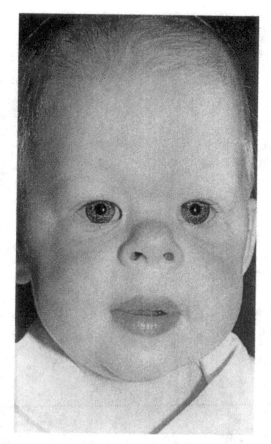

FIG 17.7—*The typical facial features of Williams' syndrome. The face is described as "elfin", the nostrils are anteverted, the philtrum is long, and there is relative midface hypoplasia. Often the iris has a stellate pattern. (Reproduced by kind permission of Dr Michael Baraitser and Professor Robin Winter.)*

specifically he pulls his nails out and puts various objects into his mouth and anus causing trauma. These last features had required acute medical treatment and led to the patient being placed on an "at risk" register by social services. At the time of consultation the parents were defending themselves against charges of child abuse. In view of the behavioural characteristics, a possible diagnosis of Smith–Magenis syndrome was suggested and the diagnostic test, FISH of the short arm of chromosome 17, was undertaken. This confirmed the characteristic submicroscopic deletion of

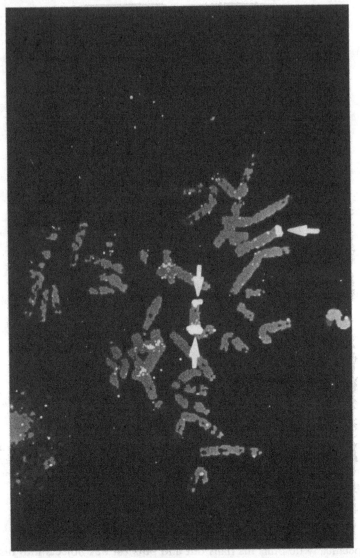

FIG 17.8—*Fluorescent in situ hybridisation (FISH) of the elastin gene in Williams' syndrome. The two chromosome 7s are identified using a chromosome 7 marker and the elastin locus is shown to be deleted in one, confirming the clinical impression of Williams' syndrome.*

chromosome 17p, thus securing the diagnosis (fig 17.10). It was now possible to refute the allegations of child abuse, self mutilation being widely reported in patients with the condition, and to counsel

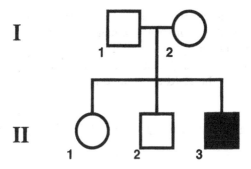

FIG 17.9—*The proband, II₁₁₁, is an isolated case in his family.*

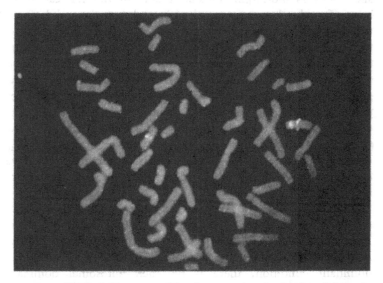

FIG 17.10—*FISH of chromosome 17p, demonstrating the deletion of the region characteristic of Smith–Magenis syndrome. Note the centromeric probe used to identify both chromosomes 17. The 17p signal is present on only one chromosome, denoting deletion of the opposite chromosome.*

the parents that, as Smith–Magenis syndrome is a sporadically occurring condition, the recurrence risk in a future pregnancy was low.

Commentary

Ten years ago this condition was unknown. It has emerged through a series of case reports and the development of molecular cytogenetics. Clearly, there is no curative therapy in these children, but securing the diagnosis frequently provides a degree of consolation and insight to families attempting to cope with behaviourally disruptive children. In the example quoted, the diagnosis led to the child abuse investigation being dropped.

Recognising new mechanisms of clinical genetic disease

In the example in figure 17·11, IV_7 developed diabetes mellitus aged 24 years and required insulin treatment. Several years later, he was found to have a progressive bilateral sensorineural deafness and died of a dilated cardiomyopathy aged 43 years. IV_4 presented with progressive hearing loss from the age of 20 years. At the age of 54 a glucose tolerance test confirmed diabetes, whilst retinal examination showed a salt and pepper retinopathy. IV_6 developed gestational diabetes aged 30. Glucose intolerance persisted after pregnancy, requiring insulin treatment. Progressive hearing loss was noted from the age of 35. On examination, at the age of 47 years the major finding was salt and pepper retinopathy. IV_{10} developed diabetes aged 46 years and was also found to have a significant sensorineural hearing impairment and pigmentary retinopathy.

The family tree (fig 17.11) showed several other individuals in the pedigree sharing one or more of these clinical problems. Although the "disease" is passed from one generation to another in generations II, III, and IV, as might occur with an autosomal dominant condition, there is no such autosomal dominant condition described and, secondly, autosomal dominant conditions rarely present so clinically varied a picture. It is striking that all the affected members of the pedigree are related through the maternal line. In no case in the family has an affected man passed the condition to his children. This observation raises the possibility that the phenotype in the pedigree might be arising from mutation of mitochondrial DNA. Apart from the nucleus, the mitochondrion is the only cell organelle with DNA and, unlike nuclear DNA which is inherited equally from both parents, mitochondrial DNA is inherited almost exclusively from the mother. Moreover, the observation of deafness and retinopathy further support the possibility that the mutation causing the disease in this family might be mitochondrial, as both of these clinical features are well

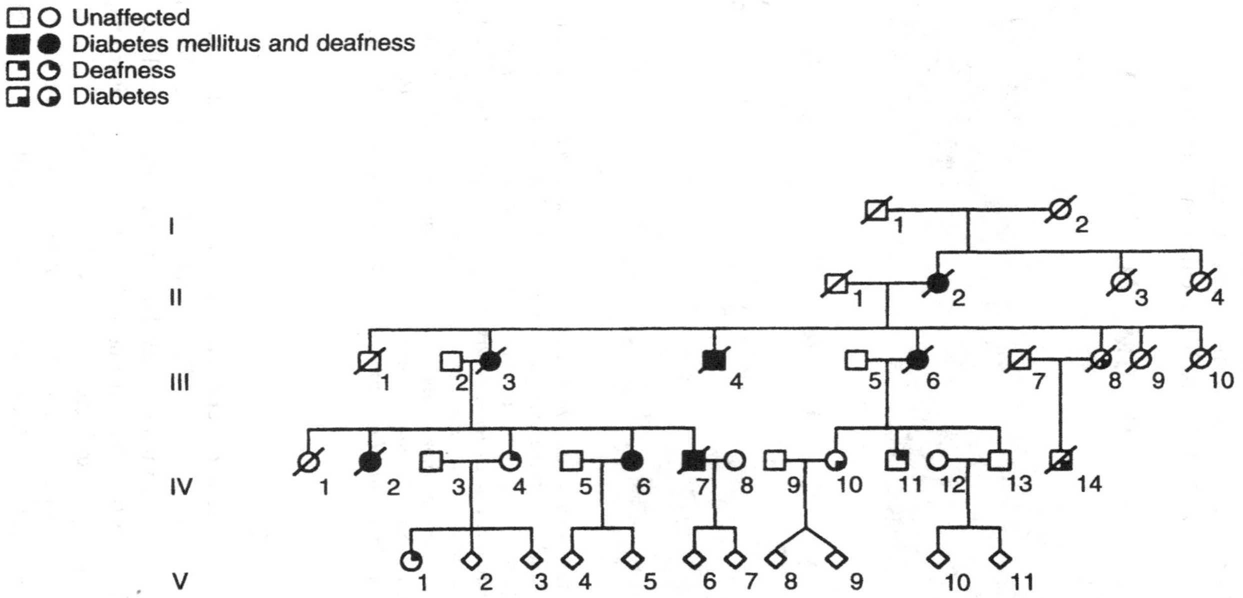
Unaffected
Diabetes mellitus and deafness
Deafness
Diabetes
I
II
III
IV
V

FIG 17.11—*The pedigree, documenting clinical features in respective patients. (Reproduced by permission of* The Lancet *from Reardon et al.*[11]*)*

described in disorders of mitochondrial dysfunction. Analysis of mitochondrial DNA confirmed this to be the case, with a mutation at position 3243 of the tRNA leucine gene (fig 17.12).[11]

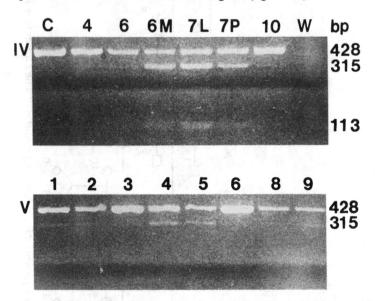

FIG 17.12—*Agarose gel photographed after amplification and cleavage with ApaI. ApaI cleaves the amplified fragment of DNA into two bands—315 and 113 base pairs long—provided that the 3243 mutation is present. In the absence of the mutation, no cleavage occurs. Samples from generation IV are shown on top. C = control, W = water; case numbers correspond to the pedigree. The DNA used was extracted from lymphocytes, except for M (muscle), L (liver), and P (pancreas). Samples from generation V below. IV$_4$, V$_3$, and V$_8$ also showed the mutation. (Reproduced by permission of The Lancet from Reardon et al.[a])*

Commentary

All the genetic material (DNA) is contained within the nucleus, with the exception of the mitochondrion, which has its own chromosome. Mutations involving the mitochondrial DNA have now been established in many different clinical disorders, which are primarily neurological in clinical presentation.[12 13] The family described is unusual in that the primary presentation is not muscle weakness, but deafness and diabetes mellitus. The mutation demonstrated in this family, involving position 3243, was first described in patients with the MELAS syndrome (mitochondrial myopathy, encephalopathy, lactic acidosis, and stroke like episodes). Neurologists have increasingly recognised these symptoms in patients with a distinct disorder of mitochondrial biochemistry. About 90% of patients presenting with this constellation of clinical signs will have the 3243 mutation, but other mutations in the mitochondrial chromosome can result in a similar phenotype.[12] Although the exact pathophysiology remains poorly understood, the observation that some families carrying the 3243 mutation may present with the alternative clinical picture of diabetes mellitus represents an important new landmark in charting the genetic contribution to the causes of diabetes. It remains unclear why the clinical consequences of the mutation may vary so widely from one family to the next.

Conclusions

The ideal practice would be to define the exact disease causing mutation in each and every family with a genetically determined condition seeking counselling. The major block to this used to be the technical difficulties in identifying disease causing genes, and the limitations that the absence of mutation detection availability placed on services to families with genetic disorders. Now that mutations for many of the more common single gene disorders have been characterised, the limitation in the provision of service often stems from the lack of availability of mutation analysis once the research funding for mutation detection expires. The transfer of research findings to clinical practice can represent the limiting step. Marfan's syndrome is a case in point. The mutation is in the fibrillin I gene on chromosome 15.[14] Although in excess of 40 mutations have now been reported in this gene in patients with Marfan's and related syndromes, almost all the observed mutations have been unique to individual families. There is not a single common mutation underlying the condition. As a result, identifying the mutation in an individual family requires dedicated laboratory staff and facilities. As this sort of resource cannot be available for each individual condition, we still need to rely on clinical parameters in deciding whether individual cases referred as possible Marfan's syndrome represent fibrillin I mutation, with consequent need for aortic arch screening, or not. As ever increasing numbers of genes are identified that are related to disease, this problem may worsen before automation in the laboratory facilities eases screening of large genes for individual mutations.

The other important trend that is already emerging is the identification of genes of known developmental significance in animal studies, but that have not been correlated with a specific disease process in humans. It is likely that many sporadically occurring malformations represent mutations of genes involved in normal cell migration and differentiation. Although it is clear from the examples cited that molecular advances have altered clinical practice in genetics, particularly for mendelian conditions, the challenge for the twenty first century is to understand the genetic basis of sporadically occurring malformations. The technological advances are either already in place or planned—it is the matching of clinical insight with basic science that is likely to be the rate limiting factor.

1 Harper PS. *Practical genetic counselling*, 4th edn. Oxford: Butterworth–Heinnemann, 1994.
2 Collins FS. Evolution of a vision: genome project origins, present and future challenges and far-reaching benefits. *Human Genome News* 1995;7(3,4):3–16.
3 McKusick VA. *Mendelian inheritance in man*, 11th edn. Baltimore: Johns Hopkins University Press, 1994.
4 Baraitser M. The genetics of neurological disorders. *Oxford monographs on medical genetics*, No. 18, 2nd edn. Oxford: Oxford University Press, 1990.
5 Hoffman EP, Brown RH, Kunkel LM, *et al*. The protein product of the Duchenne muscular dystrophy locus. *Cell* 1987;51:919–28.
6 Harper PS. *Myotonic dystrophy*, 2nd edn. London: WB Saunders, 1989.
7 Aslanidis C *et al*. Cloning of the essential myotonic dystrophy region and mapping of the putative defect. *Nature* 1992;355:548–51.
8 Buxton J *et al*. Detection of an unstable fragment of DNA specific to individuals with myotonic dystrophy. *Nature* 1992;355:547–8.
9 Harley H *et al*. Expansion of an unstable DNA region and phenotypic variation in myotonic dystrophy. *Nature* 1992;355:545–6.
10 Ewart AK *et al*. Hemizygosity at the elastin locus in a developmental disorder, Williams syndrome. *Nature Genet* 1993;5:11–16.
11 Reardon W, Ross RJM, Sweeney MG, Luxon LM, Pembrey ME, Harding AE, *et al*. Diabetes mellitus associated with a pathogenic point mutation in mitochondrial DNA. *Lancet* 1992;**340**:1376–9.
12 Hammans SR. Mitochondrial DNA-associated disease. In: Humphries SE, Malcolm S, eds. *From genotype to phenotype*. Oxford: Bios Scientific, 1994.
13 Reardon W, Harding AE. Mitochondrial genetics and deafness. *J Audiol Med* 1995;4:40–51.
14 Dietz HC. Marfan syndrome caused by a recurrent de novo missense mutation in the fibrillin gene. *Nature* 1991;352:337–9.

18: Monoclonal antibodies in medicine

Robert E Hawkins, Kerry A Chester, Meirion B Llewelyn,
Stephen J Russell

In 1975 George Köhler and César Milstein of the Medical Research
Council Laboratory of Molecular Biology in Cambridge described
an elegant system of obtaining pure antibodies of known specificities
in large amounts.[1] An astute reporter from the BBC World Service
immediately recognised the potential of the discovery of monoclonal
antibodies but it was some time before they were widely used. In
the original method, a mouse is immunised repeatedly with the
desired antigen and the spleen, which contains proliferating B cells,
is removed. B cells normally die in culture, but can be immortalised
by fusion with a non-secretory myeloma cell. The resulting
hybridoma can then secrete large amounts of the antibody encoded
by its B cell fusion partner. Supernatants from the hybrids that
survive the selection procedure are tested for binding to the original
antigen.

Alan Williams showed in 1977 that monoclonal antibodies could
be raised against biologically interesting molecules[2] and this
triggered the development of a stream of useful monoclonal based
diagnostic procedures. It soon became apparent, however, that
rodent monoclonal antibodies were unsuitable or ineffective for
treating humans because (1) they initiate human defence systems
poorly (their fine structure differs significantly from that of human
antibodies and in many cases the Fc portion is virtually unseen by
human Fc receptors and complement proteins, resulting in failure
to initiate human defence mechanisms [effector functions]), and
(2) they are themselves the target of an immune response that can
greatly shorten their circulating half life. One solution is to produce

human monoclonal antibodies, but this is difficult for several reasons. First, human B cells immortalised by fusion with a myeloma cell or by infection with Epstein–Barr virus[3] are unstable and can rapidly lose their capacity for antibody production. Even when they are successfully transformed with Epstein–Barr virus antibody yields tend to be low and the virus usually transforms IgM secreting cells that produce lower affinity antibodies. Second, hyperimmunisation of human subjects is problematic, particularly against self antigens. In vitro immunisation—that is, antigenic stimulation of cultured human lymphocytes—provides a partial solution to this problem, but antibody production is still unstable after immortalisation. The third problem is that the most convenient source of human lymphocytes is the peripheral blood, a poor source of B cells producing specific high affinity antibodies. Such cells are more abundant in the spleen, bone marrow, or lymph nodes.

Fortunately, through developments in basic sciences, solutions to these problems are emerging and the promise of human monoclonal antibody therapy is at last becoming a reality. The immunogenicity of rodent monoclonal antibody can now be reduced, to improve their ability to recruit natural effector functions and increase their affinity. Many different antibody fragments can be produced and linked to various effector functions. More recently novel methods have started to emerge for generating human antibodies directly.

Chemical modification

Regardless of how a monoclonal antibody is produced, it may be desirable to tailor it to suit its intended application better. This can be achieved through chemical or genetic approaches. Figure 18.1 shows some of the possible modifications.

Proteolytic cleavage

Controlled proteolytic cleavage of a purified monoclonal antibody gives several smaller fragments that can be separated chromatographically. Cleavage was important in the early elucidation of antibody structure and structure–function relations but also has therapeutic implications. Antibody fragments are sometimes preferable to intact antibodies as they have a shorter circulating half life and may penetrate tissues more rapidly.

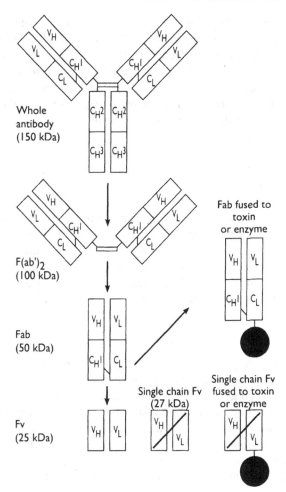

FIG 18.1—*Antibody fragments. Limited proteolytic cleavage of IgG removes the Fc portion yielding bivalent F(ab')₂ (100 kDa) or monovalent Fab antigen binding fragments. Cloned T genes for Fab can be expressed in mammalian cells, yeast, or bacteria to produce functional recombinant Fab molecules. Toxins and enzymes can be fused to the recombinant Fab by genetic engineering. Smaller antibody fragments (Fv) can be produced in bacteria by coexpression of cloned V_H and V_L genes and stability is increased when V_H and V_L are linked (genetically) by a short peptide (single chain Fv or scFv). This can also be linked to toxins genetically.*

Chemical coupling

Monoclonal antibodies and antibody fragments can be conjugated chemically to a variety of substances, including plant and bacterial toxins, enzymes, radionuclides, and cytotoxic drugs. In this way, an ineffective rodent antibody or antibody fragment may be armed with a potent effector mechanism. Fragments coupled to radioactive elements can also be used for in vitro imaging or cancer therapy. However, the chemical coupling processes can be inefficient or give rise to unstable products, and repeated cycles of antibody purification, modification, and repurification are time consuming and costly.

Genetic modification

In contrast to a hybridoma or the protein it secretes, an antibody gene is highly versatile. It can be cut, joined to other genes, mutated randomly or non-randomly, and expressed in various cell types. The genes can be introduced into appropriate plasmid vectors and transfected into mammalian cells, bacteria, insect cells, or even plant cells for protein expression.

Antibody fragments

Antibody fragments similar to those generated by proteolytic cleavage can be generated from shortened versions of the heavy and light chain genes. The modified genes can then be transfected into bacteria[4] or mammalian cells where they produce functional Fv (variable fragment) or Fab (antigen binding fragment) which is then easily purified. Bacteria are unsuitable hosts for producing complete monoclonal antibodies because the Fc domain of antibodies produced in this way is non-functional.

Antibody fusion proteins

Genetic engineering has been used to create chimaeric molecules in which the variable domain of an antibody is genetically linked to an unrelated protein. In this way enzymes, toxins, and cytokines, for example, can be given novel binding specificities and can be produced in bacteria.[5] The approach is made easier when the antibody moiety is expressed as a single chain Fv (scFv) molecule in which the heavy and light chain V domains are linked by a short peptide that does not seriously affect antigen binding. The antibody can then be expressed as a single protein from a single gene rather than as two chains that must subsequently associate somewhat inside the cell. Several bacterial toxin–scFv fusion proteins have been produced,[5] and because they are smaller than intact antibodies, it is hoped they will prove able to penetrate tumours more efficiently.

Antibodies can also be expressed on the surface of cytotoxic effector cells, redirecting them to kill novel targets. This approach has been used to redirect T cells[6] and has many potential applications to improve cellular immunotherapy.[7]

Humanisation

Chimaeric antibodies   Genetic manipulation can also be used to make chimaeric antibodies—that is, antibodies with rodent variable domains for antigen binding and human constant regions for recruiting effector functions.[8] The molecule is largely human but binds with the specificity of the parent monoclonal. Chimaerisation enhances effector functions,[9] but a significant part of the molecule is still of rodent origin and recent human trials have shown that over half of humans mount an antimouse response after receiving a chimaeric antibody.[10]

CDR grafted antibodies   Structural analysis of antibody–antigen complexes shows that the antigen binding surface of the antibody is formed by six hypervariable loops of amino acids called complementarity determining regions (CDRs). These loops are mounted on relatively constant framework regions and by genetic manipulation can be transplanted from a rodent antibody on to a human framework. This produces a CDR grafted or humanised antibody with the same specificity as the rodent monoclonal antibody from which the loops were grafted.[11] The process usually reduces the affinity of the antibody, but mutations can be made to restore full binding. Several antibodies have now been humanised[12] and one has already been used with clear therapeutic benefit.[13]

Bispecific antibodies

Bispecific antibodies have two antigen binding sites, each with a different binding specificity. Conventionally, they have been produced by fusing two hybridoma lines to make a hybrid hybridoma[14] or by chemical crosslinking of antibody fragments. As a result of random pairing of heavy and light chains and of heavy chain–light chain heterodimers, on average less than 10% of the IgG secreted by a cell expressing two antibody genes displays both of the required specificities. Genetic techniques allow the production of constructs which facilitate the association of non-identical species.[15] Bispecific monoclonal antibodies have been used to crosslink cytotoxic effector cells to targets that they would not otherwise recognise—for example, tumour cells—and the approach has been used with apparent benefit in the treatment of malignant gliomas.[16] They can also be used to redirect toxins and enzymes to specific cellular targets.

Rapid cloning of antibody genes

No two antibody genes are identical so it might be expected that cloning each gene would be a tedious process. However, with the development of rapid methods based on the polymerase chain reaction (PCR) (see chapter 3) cloning functional rearranged V genes has become routine.

Although they differ in the middle, all antibody V genes are similar at either end, which allows construction of oligonucleotide sets whose sequences recognise and bind to the terminals of most V genes and prime the polymerase chain reaction.[17] Oligonucleotide sets are available for amplification of murine or human V genes.

It is now possible to amplify and rescue most of the antibody V genes from a diverse population of human or murine B cells, thereby generating an antibody gene library. This method works equally well with any type of B cell—resting B cells,[18] antibody secreting plasma cells, or memory B cells[19]—and the starting material may be either RNA or DNA.

Whatever the source of the antibody gene library, its usefulness depends on the availability of a convenient system for expressing the genes and selecting those that encode the best antibodies. Until recently, the best system available was suitable for screening no more than a million transfected colonies of *Escherichia coli*,[20] which is at least two orders of magnitude lower than the number of antibodies screened by an intact immune system. However, with the arrival of phage antibodies (see below),[21] libraries containing at least 10^{10} different antibodies can now be screened.

Phage antibodies

Filamentous bacteriophages (hereafter referred to as phages) are pencil shaped viruses that infect bacteria. They attach to the surface of bacterial cells and inject their single stranded DNA genome through the cell wall. The infected bacterium does not die but continues to divide, distributing copies of the viral genome on to its progeny, which assemble and extrude perfect replicas of the invading phage. After overnight incubation, one millilitre of the bacterial culture supernatant contains over 10^{11} progeny phage particles.

At one tip of the phage are a few (probably three) copies of a protein (gene III protein). This protein mediates the initial attachment of the phage to a bacterial cell. To make a phage antibody the gene for an antibody fragment is fused precisely to one end of gene III on the phage genome. When this modified phage DNA is transfected into a bacterial cell, the cell produces and extrudes progeny phage particles that not only display the appropriate antibody at their tip (in fusion with the gene III protein), but also contain a single copy of the antibody gene and are still able to infect bacteria almost as efficiently as unmodified phages. Antibodies displayed on the surface of phages are fully functional and will still bind their antigen specifically. Phage antibodies with the desired specificity can be purified from a mixed population because of their ability to bind antigen.[21]

Thus a phage antibody is the functional, in vitro equivalent of a resting B cell. It contains an antibody gene and displays the corresponding functional antibody on its surface. It can be selected for its ability to recognise a particular antigen, whereupon it can be amplified by growth in bacterial culture. It is then simple to rescue the antibody gene, which can be used to produce large amounts of soluble antibody (like the plasma cell) or simply stored in the freezer (acting similarly to the memory cell) (Fig 18.2).

Phage antibody libraries

The intact humoral immune system is essentially a large library of antibodies. After challenge with antigen the most suitable antibodies are selected, amplified, and affinity matured. If the whole process could be reproduced entirely in vitro, production of high affinity human monoclonal antibodies might be greatly

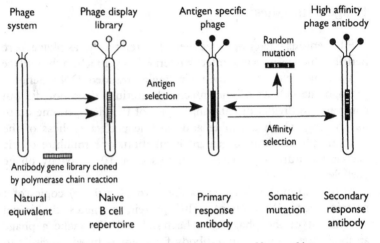

FIG 18.2—*In vitro antibodies compared with the natural humoral immune system. The phage system allows all aspects of the humoral immune system to be mimicked in vitro.*

simplified.[22] With phage antibody libraries (see below) this goal is fast becoming a reality and these should increase the pace at which antibody therapy develops.

If a phage antibody can be likened to a B cell, then a phage antibody library is the in vitro equivalent of the humoral immune systems (Fig 18.2). A phage antibody library is constructed by ligating an antibody gene library (amplified by the polymerase chain reaction) into the appropriate site on purified phage DNA. The ligated DNA is transfected into bacteria, which then manufacture large numbers of phage antibodies, and those with the desired binding specificities are selected by using soluble tagged antigen or an antigen coated surface. Phage antibody technology is changing fast. The original libraries were limited to around 10^8 in size,[18] but now techniques allow this to be increased to 10^{10} with a corresponding increase in the quality of the antibodies generated.[23]

Mice make a comeback?

More recently, it has been possible to make mice transgenic for the human immunoglobulin locus.[24] Monoclonal antibodies can be made in this way by using Köhler and Milstein's original method or, as with normal mice,[25] it may prove advantageous to make phage libraries and select for the best antibodies.[26]

Future of antibody engineering

The technological trickery of antibody engineering is advancing more rapidly than it can be tested in therapeutic models. This presents a problem for pharmaceutical companies because in the time it takes them to scale up production methods for their most promising therapeutic monoclonal antibody, both the production method and the antibody may have been superseded. Notwithstanding, several companies have taken the plunge, and monoclonal antibodies are beginning to trickle into clinical practice. This trickle will probably soon become a flood and, faced with a plethora of cleverly conceived and constructed, but competing reagents, clinicians will benefit from an understanding of some of the principles of antibody therapy.

The 1990s will be a testing time for monoclonal antibodies. Potential clinical applications include the treatment of cancer, autoimmune disease, transplant rejection, viral infection, and toxic shock. The Centre for Exploitation of Science and Technology has estimated that the total world market for monoclonal antibodies will have reached $1000 million by 1994, rising to $6000 million by the year 2000.[27] It remains to be seen whether the clinical promise of monoclonal antibodies will be realised on such a grand scale, but antibody therapy is likely to be much in evidence in many clinical settings over the next few years. Clinicians will therefore need to familiarise themselves with some of the issues relating to use of clinical antibodies.

Target antigens

Antibodies can neutralise toxins, block the interaction of growth factors, hormones, intercellular adhesion molecules, or viruses with their cognate cellular receptors, and coat bacteria, viruses, or cells, marking them for phagocytosis, antibody dependent cellular cytotoxicity, or complement mediated lysis.

Target antigens can therefore be circulating or on the cell surface. Selecting a suitable target for a given disease depends not only on the aims of treatment, but also on the precise tissue distribution of the target antigen, its function, and its fate after it has complexed with the therapeutic monoclonal antibody. Finding a suitable target antigen is probably the most important factor determining the ultimate success or failure of antibody therapy. Provided that the target has been well chosen it may be possible to modify the corresponding monoclonal antibody in various ways to enhance its therapeutic potential.

Pharmacokinetics

Infused antibodies are diluted almost immediately in the total plasma volume and then diffuse more slowly across the walls of small blood vessels into the interstitial fluid (distribution phase). The half life of the circulating antibody is determined by the rate at which it is metabolised and excreted (elimination phase). The degree to which the target antigen is bound varies with the total time of exposure, the concentration, and the kinetic properties of the monoclonal antibody.

Reaching the target

How easily infused monoclonal antibody can reach it depends on the target's location. Intravascular targets are readily accessible, but other targets are less easily reached because exit from the vascular system is restricted. To gain access to extravascular targets (for example, cancer cell surface antigens) the antibody must pass through the endothelial lining of a capillary or postcapillary venule. Smaller antibody fragments, particularly Fv reagents, penetrate the interstitial fluid space more readily than whole IgG.[28] High molecular weight proteins can escape from the microvessels through gaps between adjacent endothelial cells, which are particularly abundant in inflamed tissues. The discontinuous endothelial lining of the sinusoidal circulations of liver, spleen, and bone marrow also allows free passage of macromolecules such as IgG.

Binding reaction

The degree to which a therapeutic antibody binds to its target antigen is governed by the concentration of antigen, the concentration of antibody to which it is exposed, the duration of exposure, and the intrinsic properties of antibody and antigen that determine their rates of association and dissociation.

The equation is simple when attempting, for example, to neutralise a circulating toxin. For cell surface antigens the analysis is less simple. The concentration of antibody to which the target cells are exposed and the duration of exposure are determined by the rate at which the antibody enters the interstitial fluid and the rate at which it is eliminated from the body. Also, cell associated targets are effectively multivalent so that affinity is no longer the only factor determining the rate of dissociation of cell bound antibody. Provided that the target antigen is expressed at sufficiently high density, bivalent molecules such as intact IgG or F(ab)$_2$ can bind with much greater avidity than can the smaller univalent Fab and Fv antibody fragments. Complexed antibody may also be taken into the cytoplasm of the target cell, effectively preventing further dissociation. Additionally, a cluster of many cells displaying the same target antigen (a tumour deposit, for example) may behave as an antigen "sink" from which the antibody escapes only very slowly. This is because dissociation of a bound antibody molecule will be followed immediately by rebinding to the same or a neighbouring cell.

Antibody clearance

Ultimately, all infused antibody will be eliminated from the body. The Fc portion of IgG is thought to determine its catabolic rate, which (in humans) is faster for murine antibodies than for human antibodies.[29] Smaller antibody fragments pass relatively easily from the glomerular capillaries into the renal tubules and are rapidly excreted unchanged in the urine; this greatly shortens their plasma half life. Thus the circulating half life of IgG is measured in days, and that of single chain Fv fragments (scFv) in minutes, while F(ab)$_2$ and Fab fragments have intermediate half lives.[28]

Clearance of antibody that has been retained in the tissues is slower. Retention of antibodies by tissue because of specific interaction of the antibody with its target antigen is welcome, but non-specific binding to homologous or non-homologous antigens also occurs. Moreover, IgG may be retained in liver, spleen, and bone marrow through the interaction of its Fc portion with Fc receptors on resident macrophages. Fab and F(ab)$_2$ fragments tend to accumulate in the kidneys.[29] Persistence of antibodies in normal host tissues may be troublesome, leading, for example, to excessive toxicity of a radiolabelled therapeutic cancer antibody.

Effector mechanisms

When the goal of treatment is neutralisation of a toxin or blockade of a ligand–receptor interaction the therapeutic antibody requires no special effector domain and, depending on the effective valency of the target antigen, should function well as a monovalent (single chain Fv or Fab) or bivalent (F(ab)$_2$) fragment. More commonly, however, the aim is to destroy a specific population of target cells. Phagocytosis, antibody dependent cellular cytotoxicity, and complement fixation are the natural effector pathways activated by the Fc portion of cell bound antibody. Smaller antibody fragments without an Fc portion (single chain Fv, Fab, F(ab)$_2$) can be artificially given alternative effector mechanisms including radioactive metals, plant and bacterial toxins, enzymes, and cytotoxic drugs.

The most suitable effector mechanism depends on several factors. A high density of IgG on the target cell is required to activate complement because it is initiated by crosslinking the Fc portion of two adjacent cell bound antibody molecules. Moreover, Fc mediated recruitment of phagocytes, antibody dependent cellular cytoxicity, and complement is not possible unless bound antibody stays on the surface of the target cells.

For some conditions it may be more appropriate to use antibodies with artificially linked effector functions. Cells that rapidly internalise bound antibody or express the target antigen at low density may be killed more effectively by antibodies conjugated to drugs, toxins, or radionuclides. Radioimmunoconjugates also have the advantage that their radiation can penetrate several cell diameters into the tissues—this may be useful for cancer therapy as the monoclonal antibody cannot penetrate deep into the tumour. The box summarises the available antibody effector mechanisms.

Clinical use of monoclonal antibodies

Immunosuppression

Monoclonal antibodies offer a realistic alternative to immunosuppressive drugs, and this is perhaps their most useful current application. Potential targets for immunosuppressive monoclonal antibodies include lymphocyte differentiation antigens, cytokines, cytokine receptors, and cell adhesion molecules.

Effector functions for antibody targeted therapy

Blocking   Ligand–receptor interactions
- Cell adhesion
- Virus attachment
- Cytokine stimulation

Natural   Fc mediated
- Complement fixation
- Antibody dependent cellular cytotoxicity
- Phagocytosis

Artificial   Toxins
- Plant toxins
- Bacterial toxins

Radioisotopes

Cytotoxic drugs

Enzymes
- Prodrug activation
- Direct toxicity

Bifunctional   Crosslinking
- Cytotoxic effectors to targets
- Enzymes or toxins to targets

The first monoclonal antibody to be approved for human therapy (OKT3) is an immunosuppressive murine reagent which binds to T lymphocytes and is useful for treating rejection of renal transplants.[30][31] In common with many other immunosuppressive antilymphocyte monoclonal antibodies it does not stimulate a strong antimouse response. The toxicity of OKT3 is worse with the first dose, which triggers release of cytokines from targeted cells and leads in some cases to hypotension, weight gain, and breathlessness, progressing occasionally to pulmonary oedema. Many other immunosuppressive monoclonal antibodies have been shown to have activity in humans. Among the most promising are antibodies against the lymphocyte antigens CD4, CD28, Tac, and CDw52 (see below), all of which have now been humanised by CDR grafting,[32-34] and several monoclonal antibodies that block adhesion of immune and inflammatory cells.

Monoclonal antibodies against CD4 inhibit the function of helper T cells and have been used with varying success to treat acute rejection of renal allografts, rheumatoid arthritis, inflammatory bowel disease, systemic lupus erythematosus, psoriasis, relapsing polychondritis, systemic vasculitis, and mycosis fungoides.[35-41] Tac monoclonal antibodies recognise high affinity interleukin-2 receptors of activated lymphocytes and do not bind to resting lymphocytes. They can therefore block ongoing antigen specific immune responses highly specifically without damaging resting lymphocytes. Murine Tac monoclonal antibodies were shown to prevent early rejection of renal allografts, but antimouse responses were detected in 81% of patients after one month of treatment.[42] Humanised Tac antibody (Tac-H) was recently compared with the murine antibody in primates given cardiac allografts.[43] The humanised antibody had a longer circulating half life (103 vs 38 h), was less immunogenic (0% vs 100% antiantibody responses before day 33), and produced a longer graft survival than the murine antibody.

Immunosuppressive monoclonal antibodies will undoubtedly contribute to the therapeutic options against autoimmune disease and rejection of transplants but their precise role has yet to be defined. More detailed understanding of the underlying immunopathogenic mechanisms in many autoimmune conditions and vasculitic states will help future exploration of the therapeutic potential of monoclonal antibodies.

Infection

Agammaglobulinaemic patients suffer from recurrent bacterial sinopulmonary infection, meningitis, and bacteraemia.[44] Viral infections are no more severe in patients with agammaglobulinaemia than in healthy people, suggesting that T cells are the most important initial defence, but lasting immunity is lacking so multiple bouts of chickenpox and measles may occur. These observations suggest that antibodies should be able to prevent bacterial and viral infections. Indeed, regular administration of purified pooled human immunoglobulin provides good protection for patients with agammaglobulinaemia (and hypogammaglobulinaemia or dysgammaglobulinaemia).[45]

Polyclonal human immunoglobulin preparations have been used for many years to treat and prevent several viral diseases including hepatitis A and B, chickenpox, measles, and cytomegalovirus infection. Several antiviral and antibacterial monoclonal antibodies are now under development for human risks. For example, humanised versions of monoclonal antibodies to herpes simplex virus[46] and respiratory syncytial virus[47] have been prepared and human antibodies to HIV have been isolated by screening phage libraries.[48] Antiviral monoclonal antibodies can block attachment and penetration of viruses, opsonise virus and virus infected cells for phagocytosis or antibody dependent cytoxicity, and mediate complement lysis of enveloped virus particles or infected cells. Cocktails of monoclonal antibodies will probably give greater benefit than single reagents.

However, as T cells and not antibodies seem to be essential for eradicating established viral infections, it can be argued that antibodies are unlikely to be useful in treating these conditions. Moreover, there is evidence that certain viral infections may be enhanced by antiviral antibodies, which can facilitate Fc receptor mediated viral entry into macrophages and some other cells.

Toxic states

The use of monoclonal antibodies to treat septic shock has been reviewed recently.[49] Endotoxin, a lipopolysaccharide component of the bacterial cell wall, damages vascular endothelium, which triggers a cascade of events that leads to septic shock. Because the target antigen is intravascular, IgM monoclonal antibodies can be used. HA-1A, a human IgM anti-endotoxin monoclonal antibody, reduced 28 day mortality by 39% in 105 patients with Gram negative bacteraemia.[50] Although the result seems impressive, the antibody is expensive and it is difficult to design protocols that avoid treating large numbers of patients who subsequently prove not to have had Gram negative bacteraemia. Moreover, the initial study has been seriously criticised and a second placebo controlled clinical trial of HA-1A has been recommended to assess whether the antibody should be widely used.[51]

Tumour necrosis factor is one of the central mediators of septic shock, and monoclonal antibodies against it are protective in animal models. A phase I clinical trial of one monoclonal antibody against tumour necrosis factor confirmed its safety, but its efficacy has yet to be shown in humans.[52] Besides the obvious examples of tetanus and diphtheria, other toxic states that may be amenable to monoclonal antibody therapy include drug overdosage, chemical poisoning, and snake or spider bites. Already digoxin Fab fragments are well established for the management of digoxin overdose and monoclonal antibodies are being developed for neutralising tricyclic antidepressants.[53]

Cancer (solid tumours)

Monoclonal antibodies against cancers have been used for both imaging and treatment. There are numerous possible target antigens, which fall into several broad categories (see box). Except in a few cases, unique tumour specific antigens have not been identified, and studies have focused on target antigens that are present to a greater or lesser degree on some normal host tissues. Examples include oncofetal antigens such as carcinoembryonic antigen and α fetoprotein, epidermal growth factor receptors, carbohydrate antigens, and components of the extracellular matrix such as mucin. Radioimmunoconjugates accumulate in tumour deposits well enough to produce reasonable images,[54] although the image is not yet good enough seriously to challenge conventional imaging methods such as computed tomography. Treatment of advanced cancer with monoclonal antibodies has so far been disappointing.[55] Early studies used immunogenic murine monoclonal antibodies that could not recruit human effector functions. Humanising these monoclonal antibodies or linking them to radioisotopes, toxins, and drugs (which may increase immunogenicity) has so far had little impact on their therapeutic efficacy and can produce serious toxicities. However, it would be inappropriate to discount the potential of these alternative killing mechanisms.

An important limiting factor is the inability of infused monoclonal antibodies to reach the target cells. Monoclonal antibodies have good access to the tumour surface as the surface blood vessels of a tumour deposit are relatively leaky to macromolecules but the branches of these vessels, which penetrate the tumour parenchyma, are not.[55 56] Once on the surface, however, they meet an impenetrable wall of tumour cells held together by tight intercellular junctions, which makes access to deeper parenchymal regions of the tumour poor. It was hoped that smaller versions of the antibody molecule—for example, single chain Fv—would escape more readily from penetrating vessels and permeate better through the parenchymal regions of the tumour. However, the early signs are that their lack of avidity (they are univalent) and rapid renal excretion result in lower absolute tumour uptake despite a better

---

Target antigens for monoclonal antibodies against cancer

Unique to tumour	Immunoglobulins
	T cell receptors
	Mutated cell surface proteins
Relative abundance in tumour	Growth factor receptors
	Oncofetal antigens
	Dead cell markers
Confined to tumour and	Differentiation antigens
non-essential normal tissues	
Stromal targets	Endothelial activation markers
	Fibroblast activation markers

---

tumour to normal tissue ratio.[28] However, it remains to be seen if scFv developed using phage technology will be superior in this respect.[25]

Problems of access are less severe in patients with low tumour burden. It is notable that the only major trial *of antibody* therapy in an adjuvant setting produced survival benefits comparable to those obtained with chemotherapy, but with less toxicity.[57] This appears a promising way forward.

The toxicity of cancer monoclonal antibodies has been variable. With unmodified murine monoclonal antibodies fever, rigors, nausea, and vomiting are common after the initial doses, immediate hypersensitivity reactions can occur, and symptoms secondary to circulating immune complexes are sometimes seen after prolonged treatment. Radioimmunoconjugates usually cause appreciable toxicity to normal bone marrow, and immunotoxins can cause the vascular leak syndrome.

Antibody dependent enzyme prodrug therapy (ADEPT) is a promising research prospect.[58] With this technique an antibody–enzyme conjugate is administered, which localises to tumour deposits. After a few days, during which non-specifically bound monoclonal antibody is cleared, an inactive prodrug is administered. The prodrug is converted by monoclonal antibody linked enzyme in tumour deposits to an active tumoricidal drug that is small enough to permeate the deeper regions of the tumour. Human and humanised antibodies with improved affinity and specificity are likely to be used increasingly in the future. One recent animal experiment has shown that improved affinity can give improved anticancer activity.[59] Phage technology could help develop appropriate antibodies.[26] Cocktails of monoclonal antibodies may give better results than single antibodies (see below).

Haematological malignancies

Monoclonal antibody treatment for haematological malignancies has been more successful than that for solid tumours. Activity against disease in bone marrow and spleen has been notable, with nodal disease responding less readily.[60] One possible explanation is that the sinusoidal circulations of responsive organs are easily permeated by immunoglobulins. Also the bone marrow and spleen are rich in host effector cells that can recognise and kill targets coated by monoclonal antibodies.

Murine anti-idiotypic monoclonal antibodies have been raised against unique surface immunoglobulins and T cell receptors expressed respectively on B and T cell malignancies. The results of treatment are encouraging[61 62] but monoclonal antibodies have to be tailor made for each patient. Alternative targets include a wide array of leucocyte differentiation antigens. B cell malignancies, for example, have been treated with monoclonal antibodies against the lymphocyte antigens CD19, CD22, CD37, and CDw52.[60 63 64] The monoclonal antibodies inevitably destroy some normal lymphocytes, but these are regenerated from stem cells, which are not attacked. Transient antibody related immunosuppression can, however, be troublesome.

Natural effector functions are effective against several haematological malignancies. CAMPATH-1G is a rat IgG2b monoclonal antibody that recruits human complement and antibody dependent cytotoxicity and binds to an antigen (CDw52) present on most normal and malignant lymphocytes. Of 29 patients with lymphoid malignancies who received CAMPATH-1G, nine attained complete remissions, although disease in lymph nodes was generally resistant to treatment.[60] A CDR grafted version of this antibody (CAMPATH-1H) was the first humanised monoclonal antibody to enter clinical trials and induced complete remissions in two patients with B cell non-Hodgkin's lymphoma, one with lymph node disease.[13] The results of larger scale trials are yet to be reported.

Immunotoxins[63] and radioimmunoconjugates[64] have shown activity against lymphoma, and certainly high dose radioimmunotherapy can be very effective, although there is also considerable toxicity.[65] Polyclonal antiferritin antisera have been shown to target Hodgkin's disease deposits more efficiently than antiferritin monoclonal antibodies,[66] which suggests again that monoclonal antibody cocktails may be the best way forward.

Other applications

Monoclonal antibodies are being developed for imaging of infarcted myocardium (antimyosin), deep venous or arterial thromboses (antifibrin), and foci of infection of inflammation. Antirhesus monoclonal antibodies have been made for treating rhesus haemolytic disease and antiplatelet monoclonal antibodies for prevention of intravascular thrombosis. Monoclonal antibody enzyme conjugates targeted at blood clots are also under development as novel fibrinolytic reagents.

Conclusions

We have progressed considerably since the early days of monoclonal antibody therapy but there is still much to learn. Human (or humanised) monoclonal antibodies are preferable to rodent monoclonal antibodies for most applications. Cocktails of monoclonal antibodies should be more effective than single antibodies, and production of such cocktails will be helped by the advent of human phage antibody libraries. Enhancement of an antibody's affinity is now possible by phage technology, and the early signs suggest that it should improve therapeutic efficacy. Definitive studies comparing the clinical efficacy of various natural and artificial effector functions are needed, and there is scope for boosting natural effector mechanisms with lymphokine therapy. For the future, antibodies and antibody genes may be used increasingly to redirect cytotoxic cells or for targeted delivery of genes and other drugs wrapped up in viruses or liposomes.

1 Kohler G, Milstein C. Continuous cultures of fused cells secreting antibody of predefined specificity. *Nature* 1975;**256**:495–7.
2 Williams AF, Galfre G, Milstein C. Analysis of cell surfaces by xenogenic myeloma-hybrid antibodies: Differentiation antigens of rat lymphocytes. *Cell* 1977;**12**:663–73.
3 Steinitz M, Izak G, Cohen S, Ehrenfeld M, Flechner I. Continuous production of monoclonal rheumatoid factor by EBV-transformed lymphocytes. *Nature* 1980;**287**:443–5.

4 Skerra A, Pluckthün A. Assembly of a functional immunoglobulin Fv fragment in Escherichia coli. *Science* 1988;**240**:1038–41.

5 Pastan I, Fitzgerald D. Recombinant toxins for cancer treatment. *Science* 1991; **254**:1173–7.

6 Gross G, Waks T, Eshhar Z. Expression of immunoglobulin-T-cell receptor chimeric molecules as functional receptors with antibody-type specificity. *Proc Natl Acad Sci USA* 1989;**86**:10024–8.

7 Rosenburg SA. The immunotherapy and gene therapy of cancer. *J Clin Oncol* 1992;**10**:180–99.

8 Neuberger MS. Williams GT, Mitchell EB, Jouhal SS, Flanagan JG, Rabbitts TH. A hapten-specific chimaeric IgE with human physiological effector function. *Nature* 1985;**314**:268–70.

9 Brüggemann M, Williams GT, Bindon CI, Clark MR, Walker MR, Jefferis R, *et al.* Comparison of the effector functions of human immunoglobulins using a matched set of chimeric antibodies. *J Exp Med* 1987;**166**:1351–61.

10 Meredith RF, Khazaeli MB, Plot WE, Saleh MN, Liu T, Allen LF, *et al.* Phase I trial of iodine-131-chimeric B72.3 (human IgG4) in metastatic colorectal cancer. *J Nucl Med* 1992;**33**:23–9.

11 Jones PT, Dear PH, Foote J, Neuberger MS, Winter G. Replacing the complementarity-determining regions of a human antibody with those from a mouse. *Nature* 1986;**321**:522–5.

12 Russell SJ, Llewelyn MB, Hawkins RE. The human antibody library: entering the next phage. *BMJ* 1992;**304**:585–6.

13 Hale G, Clark MR, Marcus R, Winter G, Dyer MJS, Phillips JM, *et al.* Remission induction in non-Hodgkin lymphoma with reshaped monoclonal antibody CAMPATH-1H. *Lancet* 1988;**ii**:1394–9.

14 Milstein C, Cuello AC. Hybrid hybridomas and their use in immunohistochemistry. *Nature* 1983;**305**:537–40.

15 Kostelny SA, Cole MS, Tso JY. Formation of a bispecific antibody by use of leucine zippers. *J Immunol* 1992;**148**:1547–53.

16 Nitta T, Sato K, Yagita H, Okumura K, Ishi S. Preliminary trial of specific targeting therapy against malignant glioma. *Lancet* 1990;**335**:368–71.

17 Orlandi R, Güssow DH, Jones PT, Winter G. Cloning immunoglobulin variable domains for expression by the polymerase chain reaction. *Proc Natl Acad Sci USA* 1989;**86**:3833–7.

18 Marks JD, Hoogenboom HR, Bonnett TP, MacCafferty J, Griffiths AD, Winter G. By-passing immunization: human antibodies from V-gene libraries displayed on bacteriophage. *J Mol Biol* 1991;**222**:581–97.

19 Hawkins RE, Winter G. Cell selection strategies for making antibodies from variable gene libraries: trapping the memory pool. *Eur J Immunol* 1992;**22**: 867–70.

20 Huse WD, Sastry L, Iverson S, Kang AS, Alting-Mees M, Burton DR, *et al.* Generation of a large combinatorial library of the immunoglobulin library in phage lambda. *Science* 1989;**246**:1275–81.

21 MacCafferty J, Griffiths AD, Winter G, Chiswell DJ. Phage antibodies: filamentous phage displaying antibody variable domains. *Nature* 1990;**348**: 552–4.

22 Winter G, Milstein C. Man-made antibodies. *Nature* 1991;**349**:293–9.

23 Griffiths AD, Williams SC, Hartley O, Tomlinson IM, Waterhouse P, Crosby WL, *et al.* Isolation of high affinity human antibodies directly from large synthetic repertoires. *EMBO J* 1994;**13**:3245–60.

24 Lonberg N, Taylor LD, Harding FA, Trounstine M, Higgins KM, Schramm SR, *et al.* Antigen-specific human antibodies from mice comprising four distinct genetic modifications. *Nature* 1994;**368**:856–9.

25 Chester KA, Bergent RHJ, Robson L, Keep P, Pedley RB, Boder JA, *et al.* Phage libraries for generation of clinically useful antibodies. *Lancet* 1994;**343**: 455–6.

26 Hawkins RE, Russell SJ, Winter G. Selection of phage antibodies by binding affinity: mimicking affinity maturation. *J Mol Biol* 1992;**226**:889–96.

27 Savin J. *The value of antibody engineering technology to the UK*. London: Centre for Exploitation of Science and Technology, 1990.

28 Milenic DE, Yokota T, Filpula DR, Finkelman MAJ, Dodd SW, Wood JF, *et al*. Construction, binding properties, metabolism, and tumor targeting of a single-chain Fv derived from the pancarcinoma monoclonal antibody CC49. *Cancer Res* 1991;**51**:6363–71.

29 Waldmann TA. Monoclonal antibodies in diagnosis and therapy. *Science* 1991; **252**:1657–62.

30 Ortho Multicentre Transplant Study Group. A randomised trial of OKT3 monoclonal antibody for acute rejection of cadaveric renal transplants. *N Engl J Med* 1985;**13**:337–42.

31 Carpenter CB. Immunosuppression in organ transplantation. *N Engl J Med* 1990;**322**:1224–6.

32 Gorman SD, Clark MR, Routledge EG, Cobbold SP, Waldmann H. Reshaping a therapeutic CD4 antibody. *Proc Natl Acad Sci USA* 1991;**88**:4181–5.

33 Queen C, Schneider WP, Selick HE, Payne PW, Landolfi NF, Duncan JF, *et al*. A humanized antibody that binds to the interleukin 2 receptor. *Proc Natl Acad Sci USA* 1989;**86**:10029–33.

34 Riechmann L, Clark M, Waldmann H, Winter G. Reshaping human antibodies for therapy. *Nature* 1988;**332**:323–7.

35 Reinke P, Miller H, Fietze E, Herberger D, Volk HD, Neuhaus K, *et al*. Anti-CD4 therapy of acute rejection in long-term renal allograft recipients. *Lancet* 1991;**338**:702–3.

36 Horneff G, Burmester GR, Emmrich F, Kalden JR. Treatment of rheumatoid arthritis with an anti-CD4 monoclonal antibody. *Arthritis Rheum* 1991;**34**: 129–40.

37 Emmrich J, Seyfarth M, Fleig WE, Emmrich F. Treatment of inflammatory bowel disease with anti-CD4 monoclonal antibody. *Lancet* 1991;**338**:570–1.

38 Hiepe F, Volk HD, Apostoloff E, Baehr RV, Emmrich F. Treatment of severe systemic lupus erythematosus with anti-CD4 monoclonal antibody. *Lancet* 1991; **338**:1529–30.

39 Nicolas JF, Chamchick N, Thivolet J, Wijdenes NB, Morel P, Revillard JP. CD4 antibody treatment to severe psoriasis. *Lancet* 1991;**338**:321.

40 Van der Lubbe PA, Miltenburg AM, Breedveld FC. Anti-CD4 monoclonal antibody for relapsing polychondritis. *Lancet* 1991;**337**:1349.

41 Mathieson PW, Cobbold SP, Hale G, Clark MR, Oliveira DBG, Lockwood CM, *et al*. Monoclonal antibody therapy in systemic vasculitis. *N Engl J Med* 1990;**323**:250–4.

42 Sollilou J, Cantarovich D, LeMauff B, Giral M, Robillard N, Hourmant M, *et al*. Randomised controlled trial of monoclonal antibody against the interleukin 2 receptor (33B3.1) as compared with rabbit antithymocyte globulin for prophylaxis against rejection of renal allografts. *N Engl J Med* 1990;**322**:1175–82.

43 Brown PS, Parenteau GL, Dirbas FM, Garsia RJ, Goldman CK, Bukowski MA, *et al*. Anti-Tac-H, a humanized antibody to the interleukin 2 receptor, prolongs primate cardiac allograft survival. *Proc Natl Acad Sci USA* 1991;**88**: 2663–7.

44 Spickett GP, Misbah SA, Chapel HM. Primary antibody deficiency in adults. *Lancet* 1991;**337**:281–4.

45 Webster ADB. Intravenous immunoglobulins. *BMJ* 1991;**303**:375–6.

46 Co MS, Deschamps M, Whitley RJ, Queen C. Humanized antibodies for antiviral therapy. *Proc Natl Acad Sci USA* 1991;**88**:2869–73.

47 Tempest PR, Bremmer P, Lambert M, Taylor G, Furze JM, Karr FJ, *et al*. Reshaping a human monoclonal antibody to inhibit human respiratory syncytial virus infection in vivo. *Biological Technology* 1991;**9**:266–71.

48 Burton DR, Barbas III CF, Persson MAA, Koenig S, Chanock RM, Lerner RA. A large array of human monoclonal antibodies to type I human immunodeficiency virus from combinatorial libraries of asymptomatic seropositive individuals. *Proc Natl Acad Sci USA* 1991;**88**:10134–7.

49 Hinds CJ. Monoclonal antibodies in sepsis and septic shock. *BMJ* 1992;**304**: 132–3.

50 Ziegler EJ, Fisher CJ Jr, Sprung CL, Straube RC, Sadoff JC, Foulke GE, *et al.* Treatment of Gram-negative bacteremia and septic shock with HA-1A human monoclonal antibody against endotoxin: a randomized, double blind, placebo-controlled trial. *N Engl J Med* 1991;**324**:429–36.

51 Warren HS, Danner RL, Munford RS. Sounding board. Anti-endotoxin monoclonal antibodies. *N Engl J Med* 1992;**326**:1153–7.

52 Exley AR, Cohen J, Buurman W, Owen R, Hanson G, Lunley J, *et al.* Monoclonal antibody to TNF in septic shock. *Lancet* 1990;**335**:1275–6.

53 Kulig K. Initial management of ingestions of toxic substances. *N Engl J Med* 1992;**326**:1677–81.

54 Order SE. Presidential address: systemic radiotherapy—the new frontier. *Int J Radiat Oncol Biol Phys* 1990;**18**:981–92.

55 Dvorak HF, Nagy JA, Dvorak AM. Structure of solid tumors and their vasculature: implications for therapy with monoclonal antibodies. *Cancer Cells* 1991;**3**:77–85.

56 Dvorak HF, Nagy JA, Dvorak JT, Dvorak AMI. Identification and characterization of the blood vessels of solid tumors that are leaky to circulating macromolecules. *Am J Pathol* 1988;**133**:95–109.

57 Rietmuller G, Schneider-Gadicke E, Schlimok G, Schmiegel W, Raab R, Hoffken K, *et al.* Randomised trial of monoclonal antibody for adjuvant therapy of resected Duke's C colorectal carcinoma. *Lancet* 1994;**343**:1177–83.

58 Bagshawe KD. Towards generating cytotoxic agents at cancer sites. *Br J Cancer* 1989;**60**:275–81.

59 Schlom J, Eggensperger D, Colcher D, Molinolo A, Houchens D, Miller LS, *et al.* Therapeutic advantage of high-affinity anticarcinoma radioimmunoconjugates. *Cancer Res* 1992;**52**:1067–72.

60 Dyer MJS, Hale G, Marcus R, Waldmann H. Remission induction in patients with lymphoid malignancies using unconjugated CAMPATH–1 monoclonal antibodies. *Leukemia Lymphoma* 1990;**2**:179–93.

61 Brown SL, Miller RA, Horning SJ, Czerwinski D, Hart SM, McElderry R, *et al.* Treatment of B cell lymphomas with anti-idiotype antibodies alone and in combination with alpha interferon. *Blood* 1989;**73**:651–61.

62 Janson CH, Tehrani MJ, Mellstedt H, Wigzell H. Anti-idiotypic monoclonal antibody to a T cell chronic lymphatic leukaemia. *Cancer Immunol Immunother* 1989;**28**:225–32.

63 Vitetta ES, Stone M, Amlot P, Fay J, May R, Til M, *et al.* Phase I immunotoxin trial in patients with B-cell lymphoma. *Cancer Res* 1991;**51**:4052–8.

64 Press OW, Eary JF, Badger CC, Martin PJ, Appelbaum FR, Levy R, *et al.* Treatment of refractory non-Hodgkin's lymphoma with radiolabelled MB-1 (anti-CD37) antibody. *J Clin Oncol* 1989;**7**:1027–38.

65 Press OW, Eary JF, Appelbaum FR, Martin PJ, Badgir CC, Nelp WB, *et al.* Radiolabelled-antibody therapy of B-cell lymphoma with autologous marrow support. *N Engl J Med* 1993;**329**:1219–24.

66 Vriesendorp HM, Herpst JM, Germack MA, Klein JLK, Leichner PK, Loudenslager DM, *et al.* Phase I-II studies of yttrium-labelled antiferritin treatment for end-stage Hodgkin's disease, including radiation therapy oncology group 87-01. *J Clin Oncol* 1991;**9**:918–28.

19: Production and use of therapeutic agents

CR Bebbington, NH Carey

In the past 30 or 40 years, advances in cell biology, endocrinology, and immunology have led to the discovery of many proteins acting at long and short range on target cells. In addition to the well known protein hormones, a large number of growth and differentiation factors have now been discovered. These have typically been identified as activities in tissue extracts or the products of in vitro cultures and are often present in these preparations in such small amounts that purification of sufficient quantities for structural studies was often difficult or impossible. Even more remote was the prospect of testing such molecules in pure form in disease models or in the clinic.

A number of advances in molecular biology have resulted in new techniques that permit such proteins to be produced efficiently on a large scale. In essence, the processes involved require the isolation of the gene encoding the desired protein and its reintroduction into a micro-organism or a tissue culture cell line, which can be grown in large scale fermenters, in such a way that the inserted gene can be efficiently expressed to produce large quantities of protein.

All proteins, whether they are hormones (such as insulin, growth hormone, or calcitonin), enzymes (such as the blood clot forming and clot lysing components), or antibodies, have a primary structure that is determined by the gene that specifies them. However, the activity of the genes is regulated in different tissues of the body. As, for most cell types, each cell has a complete complement of the genetic material of the whole organism, it is clear that not all the genes present in any one cell are expressed. For example, myosin is only produced in muscle cells and the myosin gene is not expressed in other cell types. Thus, simply introducing a gene into a foreign cell or micro-organism does not guarantee that the product of that gene will be produced from the genetically modi-cell: the appropriate control mechanisms are also needed. Consequently, one of the first challenges to be overcome in the production of proteins from isolated genes was the identification and assembly of appropriate genetic control elements to allow efficient gene expression in the chosen host cell. As the genetic control elements are themselves encoded in DNA, it is necessary to recombine stretches of DNA sequence from different sources to allow synthesis of a given protein.

Since the early 1970s it has been possible to manipulate purified DNA in vitro with the aid of a number of specific DNA modifying enzymes and certain chemical procedures for the synthesis of small sequences of DNA. The linear molecule of DNA can be cut very precisely at particular sequences recognised by the so called restriction endonucleases and can be reassembled and joined either to other isolated DNA sequences from natural sources or to chemically synthesised DNA to produce new combinations of DNA sequences. It is also possible to introduce defined mutations or sequence changes into the DNA so that modified or completely new proteins can be synthesised under the control of the appropriate control elements. As many of the procedures resemble genetic recombination in their ability to link new combinations of genetic material, the resulting DNA is often referred to as recombinant DNA.

The expression of recombinant DNA was first achieved in the bacterium *Escherichia coli*, the organism in which many of the principles of control of gene expression had first been established by genetic analysis. Establishing expression systems for *E. coli* was also facilitated by the development of plasmids as cloning vectors for propagating foreign DNA in this organism. Plasmids are small DNA circles found in many bacterial strains and often carry genes conferring resistance to particular antibiotics. They are readily isolated from *E. coli*, manipulated in vitro to insert foreign DNA, and then reintroduced into bacteria where they will replicate along with the host cell and so propagate the foreign DNA. Bacterial expression systems typically therefore use plasmids into which the necessary DNA sequences have been inserted to regulate the expression of introduced DNA.

Subsequently, expression systems have been devised for many other types of cell, including yeast and animal cells growing in culture and even genetically engineered whole animals. It rapidly became clear that *E. coli* expression systems would not be suitable for producing all proteins and that some of the other systems available would be more appropriate for certain uses, despite the fact that *E. coli* is relatively economical to grow on a large scale and can generate very high yields of recombinant proteins. The reason that *E. coli* is not always the preferred host stems from the fact that the primary structure of the protein, determined by the DNA sequence, does not by itself define the final three dimensional structure of many proteins. Proteins are modified in a number of possible ways during or after their synthesis within the cell. For example, many proteins destined for secretion from the cell into the extracellular space or the blood serum have carbohydrate structures attached at specific amino acids in the protein by glycosylating enzymes within the cell. In many cases this carbohydrate has an important effect on the properties of the protein. It may stabilise the protein in serum or, in some cases, it may be essential for biological activity. Another common modification is the covalent linkage of two cysteine residues (a chemical oxidation reaction) to produce the so called disulphide bridge, either between two distinct protein chains or within a single protein. The correct formation of disulphide bonds is crucial to the folding of many proteins into the correct three dimensional structures for activity. An important discovery for the development of expression systems was the finding that, whereas the genetic code in the DNA is read in an identical manner by practically all cells of all organisms, the protein modifications that a cell will perform are not the same in all cell types. Thus disulphide bonds of many mammalian proteins expressed in *E. coli* are found to have formed incorrectly or not at all. As a result, the structure of the protein is compromised and the protein is found as an insoluble, inactive precipitate inside the *E. coli* cell. Similarly, bacteria are incapable of adding carbohydrate or some of the other modifications essential to mammalian proteins. Nevertheless, *E. coli* expression has proved very successful for a number of relatively small proteins that are not naturally glycosylated, such as human insulin and human growth hormone. For larger or more highly modified proteins, different host cell types must be used. Yeast cells are capable of glycosylation but the carbohydrates added are of a somewhat different chemical structure from those of mammalian cells. Even mammalian tissue culture cells do not produce completely authentic recombinant proteins in all cases because the glycosylation patterns of different tissue types vary and certain specialised modifications are carried out only in certain cell types. For instance, several of the blood clotting factors require a vitamin K dependent carboxylation of certain glutamic acid residues in the protein for activity. This modification is not carried out in all mammalian expression systems and the host cell type has to be chosen with these in mind. Whether particular modifications of a protein are important for activity or pharmacokinetics will depend on the protein, and so a number of different expression systems are used for various recombinant products.

One special technique for producing potential therapeutic agents is the monoclonal antibody technique, also called the hybridoma technique. The end result of this approach is the production of a protein, a specific antibody, in a tissue culture cell. The method of deriving the producing cell line differs from the recombinant DNA methodology described above, in that it depends on the fusion in culture of two different cells to introduce the desired gene into the production cell type. A spleen derived B lymphocyte, with a very limited lifespan in culture, containing the expressed gene for a single antibody specificity is induced to fuse with an established cell line, typically a plasmacytoma cell which is capable of indefinite cell division in culture. In this way, an antibody with a defined specificity is cloned and immortalised so that it can be produced on a large scale (see chapter 18).

Therapeutic use of recombinant protein

Protein drugs are being used in increasingly diverse therapies and to review all of the existing recombinant products is outside the scope of this chapter. Examples are therefore given in three broad areas: restoration of a defined protein defect; manipulation of the body's immune or inflammatory responses; and the use of novel engineered antibodies to target cancerous cells.

Rescue of a defined protein deficiency

Insulin The first product designed for therapeutic use made by recombinant DNA techniques was human insulin. Diabetic patients have been treated with insulin extracted from animal pancreases (mainly pig and beef) for some time. Natural human insulin has not been available principally because, in the face of the satisfactory performance of the animal products, it was not considered worthwhile to set up the complex procedures for collecting human pancreases from cadavers, even if the supply would have been adequate. Now, however, a product identical to the human protein is available through fermentation of recombinant *E. coli*. Although there has been a suspicion that some patients respond differently to this form of insulin, there is an instinctive feeling that the human form should be preferable to the animal versions. Any significant benefits are, however, only likely to be revealed in the long term.

Growth hormone   Unlike insulin, growth hormone is species specific. Patients who have hypopituitarism must, therefore, be given human growth hormone. Until recently, this came from pituitaries obtained from cadavers. Such treatment was considered satisfactory until suspicions were aroused that a small number of patients had been infected through treatment with the agent giving rise to Creutzfeldt–Jakob disease, a human spongiform encephalopathy. The natural form of human growth hormone has now been removed from the market and replaced by recombinant growth hormones from several different expression systems including *E. coli*.

Haemopoietic growth factors   The production of blood cells is regulated by a series of protein hormones which respond to a variety of external stimuli to maintain appropriate cell numbers. Thus oxygen deprivation enhances the production of erythropoietin (EPO) from renal fibroblasts which acts to stimulate the differentiation of red blood cells from bone marrow precursors (erythropoiesis). The system is unbalanced in chronic renal disease and the resulting anaemia has been a significant problem in haemodialysis patients. Recombinant EPO has been produced in cultured animal cells (a Chinese hamster ovary cell line) because appropriate glycosylation is essential for activity. It is currently indicated for use in anaemia associated with chronic renal failure, non-myeloid malignancy, and anaemia associated with azidothymidine treatment in HIV infected patients.

The two other recombinant haemopoietic growth factors currently available are produced from microbial cell fermentation. GM-CSF (granulocyte–monocyte colony stimulating factor) supports the expansion of myeloid cells including granulocytes and monocytes. Glycosylated recombinant protein, derived from yeast, is used to accelerate the myeloid cell recovery in patients with lymphoid cancer receiving autologous bone marrow transplantation. G-CSF (granulocyte colony stimulating factor), which is effective in neutropenia associated with chemotherapy, is nonglycosylated and produced from *E. coli*. Several other haemopoietic factors are in clinical development produced from a variety of expression systems.

Blood clotting and clot lysing factors Defects in the clotting cascade leading to haemophilias have been treated with factor VIII or factor IX purified from human blood. Now the risk of infection with viral agents present in donated human blood can be circumvented, at least for haempophilia A, using recombinant factor VIII produced from mammalian tissue culture cells.

Tissue plasminogen activator (tPA) is part of the natural clot lysing cascade and is effective in acute myocardial infarction. The recombinant protein, produced from mammalian (CHO) cells, has however, to compete with a number of other thrombolytic agents in the health care market. Although tPA is the natural human enzyme, extensive clinical studies have tended to indicate that a combination of a bacterial enzyme streptokinase (from a strain of *Streptococcus*) and aspirin is a suitable alternative, at least for first use. Streptokinase is considerably cheaper to manufacture than recombinant tPA. However, as streptokinase is recognised by the patient's immune system as a foreign protein, the subsequent immune response precludes the use of streptokinase in a subsequent heart attack. Consequently, recombinant tPA has an important role to play in the management of this major disease.

Manipulation of the immune system

Proteins can be used as vaccines to stimulate immune responses as well as having a number of uses in modifying ongoing immune or inflammatory responses.

Hepatitis B vaccine Hepatitis B is caused by a small DNA virus and has hitherto been impossible to control by a vaccine because the virus could not be grown in any of the normal laboratory systems used for preparing viruses. Until recently, the most promising vaccine consisted of killed virus preparations from the serum of infected individuals. This has obvious drawbacks, particularly in terms of safety.

The solution has proved to be to clone a single gene from the virus, encoding the S (surface) antigen and express this in a suitable micro-organism. The S antigen is glycosylated and, in this case, yeast has been chosen as the host organism for production. The production process has greatly improved safety for both the patients and the people responsible for the vaccine manufacture because there is no hepatitis B virus present.

Such technology therefore provides a promising approach for producing vaccines for agents which are either difficult or dangerous to grow in culture or where the vaccine itself may be a danger to the patient. Similar techniques are also being used to dissect the intricacies of the immunological responses to a number of other agents including those responsible for AIDS, malaria, schistosomiasis, and trypanosomiasis.

OKT3 monoclonal antibody Monoclonal antibodies have potential in a wide variety of indications for modulating the body's immune responses and the first antibody to be licensed for therapeutic use is a mouse antibody specific for a component of the human T cell antigen receptor (CD3), manufactured from a mouse plasmacytoma cell hybridoma. OKT3 is used to suppress T cell function in the reversal of acute liver, renal, and heart transplant rejection (see also chapter 18).

Anticancer monoclonal antibodies

The immune system has long been implicated in the body's defence against cancer, a defence that obviously goes awry when the disease becomes established. This thought has led to the proposition that stimulating or substituting for the presumed defects of the endogenous system may be a route to curing or controlling cancers.

One approach has been to obtain tumour specific antibodies, by means of the monoclonal antibody technique. In most cases, these are inadequate on their own to show significant therapeutic effects on established tumours, but attachment of a variety of radionuclides or other chemotherapeutic agents appears more promising in targeting the killing agent to the tumour and hence reducing the toxicity to the rest of the body.

Many promising candidate monoclonal antibodies have now been found. One important principle that has emerged is that truly tumour specific antigens are very rare. "New" antigens that appear on tumour cells are almost invariably found on other cells at some stage in the lifetime of the individual, either in early development or on progenitor cells within the tissue from which the tumour originates. This means that, in the development of antibody based therapies, care must be taken over dosage and biodistribution among tissues and cell types.

A further feature of antibodies derived from hybridomas is that they can themselves be recognised by the patient's immune system as a foreign protein (because the large majority of such antibodies are raised by immunising mice and the resulting murine antibody shows sequence differences when compared with its human counterpart). Thus murine antibodies injected into man induce an immune response that neutralises the effect of any subsequent injections of antibody. The techniques of recombinant DNA technology can, however, be used to address this problem because it is now possible to alter the sequence of most of the antibody, to convert it to that of a human antibody. Only the relatively small region of the antibody actually involved in determining the binding specificity of the antibody (the complementarity determining regions) need remain as mouse sequence. The results of early clinical trials indicate that such engineered humanised antibodies are indeed markedly less immunogenic than their mouse precursors and are consequently better tolerated. Similar genetic engineering can also be used to modify the sequence of the antibody more radically, for instance, to remove part of the molecule to produce a smaller protein better able to penetrate solid tumour masses. Alternatively, different protein coding sequences can be added which contribute other effector functions with the aim of improving therapeutic efficacy.

It is too early to say how successful many of these approaches will be but initial clinical results suggest that modified antibodies will in the future be among the collection of drugs used to control tumour growth (see chapter 18 for further discussion of this topic).

Limitations of the technology

Many other important products are emerging from the use of this technology in human health care. The main limitations of the approach stem from the fact that the therapeutic agents are proteins. They generally cannot be taken by mouth because they do not survive exposure to the enzymes of the gut and are not absorbed from it. They must therefore be given by injection. Research into drug delivery systems which might circumvent this problem is now actively being pursued, including, for example, slow release depot formulations and compositions that protect proteins from enzymes or assist in their penetration of mucous membranes. However, it seems that most of these products will need to be given by injection for some time to come.

A second major limitation is the high cost of producing protein drugs by these techniques. The manufacturing of proteins from large scale fermentations is still a complex and expensive technology, particularly if genetically engineered animal cells are required. The proteins currently produced are potent molecules active in low doses and their use is generally confined to severe or life threatening diseases for which more cost effective treatments are not available. For proteins needed in higher doses or for very much larger numbers of patients, cost and manufacturing capacity could be prohibitive. However, as expression systems are becoming progressively more efficient, new treatments are becoming more economical to produce.

New approaches for the future

Additional proteins are continually being identified by gene cloning and further cytokines, hormones, and growth factors will undoubtedly provide a source of drug candidates. The engineering of gene sequences to produce novel designed proteins and inclusion of chemical additions to proteins postsynthetically are also proving to be important in modifying the pharmacokinetic properties as well as the biological activities of protein drugs. Such engineered proteins will undoubtedly be increasingly exploited.

A further area of active research surrounds the possibility of delivering genes themselves, rather than their protein products, to patients. Gene therapy has the great attraction that, because of the stability and replicative potential of DNA, there is the possibility of providing a real cure for chronic and inherited diseases which might at best only be controlled by repeated administration of other kinds of drugs. However, a number of significant hurdles remain to be overcome before reliable, safe, and efficient gene delivery to chosen target tissues and appropriate regulation of gene expression can be achieved. The routine use of DNA in therapy is still some way off (see also chapter 20).

Molecular biology is, however, playing an increasingly important role in the discovery of other classes of non-protein drugs. The development of modern drugs in the past 30 years or so has depended on an understanding of the basic biology of the disease, and the quality and relevance of the model systems available to test potential drugs. It has become apparent during this period that biologically active chemical compounds achieve their effects by combining with, and hence modifying the action of, one or more specific macromolecular targets in the organism, often either an enzyme or a receptor for a natural hormone.

Many drugs have been discovered and developed without knowledge of the nature or function of the drug's target molecule. It is clear, however, that a better understanding of the structure and activity of important macromolecules will help in the future design of active drugs. The genes for many enzymes, receptors, and their activators are now being cloned and expressed to produce purified proteins in quantities sufficient for biochemical screening of large banks of chemical compounds for activity against the chosen target. Such purified proteins also permit the elucidation of their three dimensional structures by x ray crystallography and other techniques. This is greatly facilitating the study of the interactions between active compounds and their protein targets, and the more precise design of candidate drugs using computer graphic modelling.

Suggested further reading

Adair JR, Bright SM. Progress with humanised antibodies. *Exp Opin Invest Drugs* 1995;4:863–70.

Anon. Biotechnology medicines approved and under development. *Gen Eng News* 1995;15(8):12–16.

Collins MK. Gene therapy for cancer. In Latchman DS, ed. *From genetics to gene therapy.* Oxford: Bios Scientific, 1994:131–45.

Old RW, Primrose SB. *Principles of gene manipulation*, 4th edn. Oxford: Blackwell Scientific, 1989.

20: Gene therapy

Heung Chong, Richard G Vile

Progress in molecular biology techniques has made possible the development of various strategies for the transfer of exogenous genes to mammalian cells. The ability to insert genes into cells in order to correct or modify function represents a novel therapeutic approach that has broad and potentially profound implications for the treatment of numerous diseases, and promises to have a significant influence on clinical practice in the decades to come. In recent years, significant advances in this field have been accomplished in the laboratory, but, despite these encouraging results and the hype created by early clinical trials of gene therapy, the limitations of current gene transfer technology and the difficulties that impede translation from experimental models to clinical practice should be recognised and kept in perspective. Various methods that can be used to transfer exogenous genes to mammalian cells will be discussed and approaches whereby these techniques may be applied to the treatment of disease will be outlined using examples.

Gene therapy—for which diseases?

Diseases resulting from a mutation in a single gene are potentially good candidates for gene therapy. A functional copy of the gene would need to be introduced into the cell together with elements to control its expression, leading to production of the missing protein and restoration of function. However, single gene disorders such as haemoglobinopathies, which require tight regulation of expression of the inserted gene, would present a more difficult challenge for corrective gene therapy. As more genes become identified, the list of disorders that are caused by single genetic defects and which may be corrected by gene therapy will increase and, although most of these would be rare, others such as cystic fibrosis may not be uncommon.

The causes of most human diseases are, however, genetically and epigenetically multifactorial and these present a different challenge and demand more varied approaches. For instance, cancer results from accumulated mutations that affect a variety of genes controlling cell growth and differentiation and it will be almost impossible to achieve correction of all the genetics defects in every cancer cell. Strategies for the gene therapy of cancer will be discussed to illustrate how gene transfer techniques may be applied in attempts to treat a common disease of multifactorial aetiology. Approaches to apply gene therapy for the treatment of infectious diseases, including HIV infection, will also be considered.

Gene therapy: general strategies

Transfer of an exogenous gene may be performed ex vivo, where cells are first removed from the patient and an in vitro culture established. The gene is then delivered to the cells, after which the cells are re-introduced into the patient. The main disadvantage of this approach is the requirement to obtain an in vitro culture which can be a time consuming and labour intensive process.

On the other hand, the direct in vivo transfer of genes to cells in situ is conceptually more simple and would be far easier to apply in a clinical setting. Such an approach for gene transfer would be comparable to the administration of a drug, but it demands a highly efficient transfer vector which would also ideally be targeted to the appropriate cells in vivo.

How can genes be transferred to cells?

Naked plasmid DNA may be taken up by cells, but this process is inefficient and thus various methods to improve gene delivery are used. The ability of viruses to infect cells can be exploited to achieve efficient gene transfer.[1-3] The most well established and widely used vectors are those derived from retroviruses and adenoviruses.[4 5]

Retroviral vectors

The genome of retroviruses consists of two RNA strands and the life cycle includes a DNA intermediate which integrates into the host cell genome (fig 20.1). The recombinant vectors are constructed by removing the viral genes between the long terminal repeats (LTRs) and replacing these with the therapeutic gene. Expression of the inserted gene, which can be up to about 6 kilobases (kb) long, can be driven by the viral promoter in the LTR or by insertion of an exogenous promoter. To obtain infectious viral particles, the vector is introduced in plasmid DNA form into a packaging cell line.[6] Such a cell line possesses the viral genes that have been deleted from the vector and that are necessary for viral replication and encapsidation. The transcripts from these helper viral sequences cannot themselves be packaged into the viral particles because their packaging signals have previously been removed. Thus, although the viral particles that are released can infect a target cell and introduce the exogenous gene, these viral vectors are replication defective (fig 20.2). These vectors can be used for in vivo gene delivery by administration of either the producer cells or supernatant containing the viral particles.

A potential hazard is the inadvertent production of replication competent retroviruses (RCR) which may arise from recombination events between the vector and viral helper sequences. Such occurrences can be minimised by using "third generation" packaging cell lines in which the helper sequences are separated so that several recombination events are needed before an RCR can be formed.[6 7] As RCRs are potentially tumorigenic, it is important to check for their presence in supernatants from producer cell lines.

Another feature of retroviruses is that they can only infect dividing cells. This property may be usefully exploited to target delivery of genes to an actively proliferating population of tumour cells without transferring the gene to surrounding quiescent normal cells.

Adenoviral vectors

A major attraction of recombinant adenoviral vectors is the high titres that may be obtained (10^{11}/ml) which compares very favourably with retroviral vectors (up to about 10^{6}–10^{7}/ml). In addition, unlike retroviral vectors, these vectors are able to infect and transfer genes to non-dividing cells, so that it is possible to

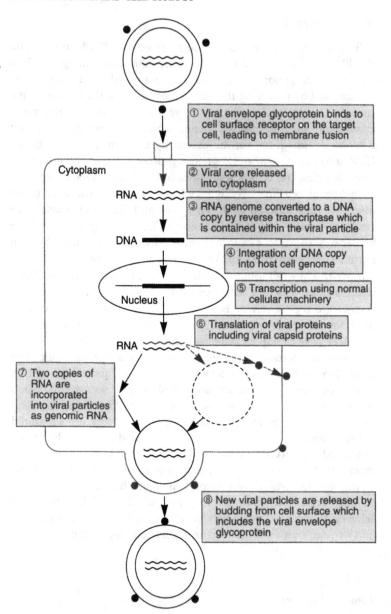

FIG 20.1—*Life cycle of retrovirus.*

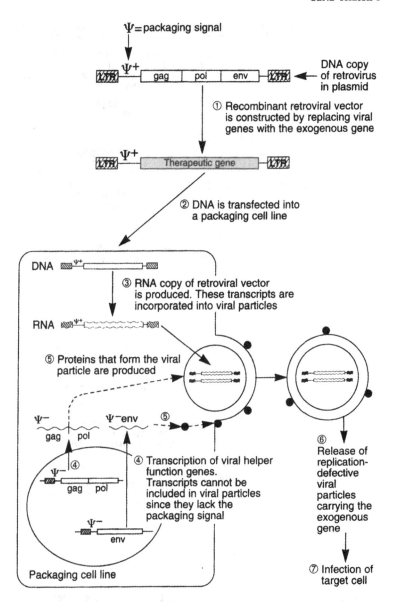

FIG 20.2—*Production of recombinant retroviral vectors.*

use them to transfer genes to terminally differentiated cells, such as bronchoepithelial cells that are targeted for gene delivery in cystic fibrosis.

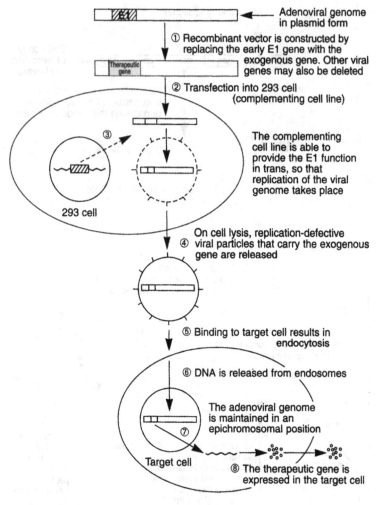

FIG 20.3—*Production of recombinant adenoviral vectors.*

Adenoviruses are DNA viruses and the recombinant vectors are constructed by deleting an early viral gene, E1, and replacing it with the exogenous gene (fig 20.3). Other viral genes may also be deleted in order to accommodate longer sequences of exogenous genes. The E1 gene is required for replication and so the vector needs to be propagated in a cell line, such as the human embryonic kidney 293 line, which is able to provide the E1 functions in *trans*.

The vector probably does not integrate into the host cell genome and thus expression of the exogenous gene cannot be expected to be long lasting. This would be a disadvantage where a permanent modification of function is desired, as would be the case for the correction of a single gene disorder, although it would not necessarily pose a problem in cancer where transient expression may be sufficient to stimulate processes that lead to cell killing. Another problem with using adenoviral vectors is the immunogenicity of the viral capsid proteins which can result in an immune response against the infected cells, thus curtailing expression of the therapeutic product, as well as priming the production of neutralising antibodies that would decrease gene transfer efficacy when the vector is re-administered. Adenoviral

---

Recombinant viral vectors

Retroviruses
Adenoviruses
Herpes simplex virus
Able to accept large inserts (up to 30 kilobases) and infect non-dividing cells. The virus can exist in a latent state in neuronal cells and thus it may be possible to develop vectors for long term gene corrective therapy for use in these cells.

Parvoviruses
Non-autonomous parvovirus: adeno associated virus
This virus has a broad host range and can probably infect non-dividing cells. The stability of the viral particle allows centrifugation to achieve very high titres. It is considered safe because it is not known to cause any human disease. Integration into host cell genome allows potential for long term expression.

Autonomous parvoviruses
Includes the prototype strain of minute virus of mice, which has an advantage of being oncotropic and hence it may be possible to construct tumour specific vectors where expression of the transgene occurs only in malignant cells.

Vaccinia virus (and other pox viruses)
This virus can accommodate large inserts (up to 25–30 kilobases). It replicates in the cytoplasm and it is possible to obtain high levels of expression of the transgene. Vaccinia vectors may be useful for expressing a tumour antigen in a "DNA vaccine" (see text).

vectors that have further deletions of viral genes are being developed. These vectors are expected to be less antigenic and thus expression of the therapeutic gene may be prolonged.

Other recombinant viral vector systems

Several other vectors based on other viruses are being developed and the features of some of these are summarised in the box on the previous page.

Non-viral vectors

A variety of non-viral methods is available for enhancing the uptake of exogenous DNA into cells (see box below).[8] These are

---

Non-viral methods of gene transfer

"Naked" plasmid DNA
Cationic liposomes
Ligand receptor mediated endocytosis
Calcium phosphate precipitation
Particle bombardment using a "gene gun"
Electroporation

---

attractive from the point of safety because there is no need to consider the inadvertent production of helper viruses. One of the most commonly used methods is to mix DNA with cationic liposomes to form complexes that have a net positive charge, thus facilitating interaction and fusion with the negatively charged cell membrane.[9] The efficiency of gene transfer is critically influenced by the ratio of DNA to liposomes in the mixture. Although viral vectors are generally considered to be more efficient than non-viral methods, successful gene transfer to multiple organs after intravenous administration of DNA–liposome complexes in animal models has been reported.[10] Another well studied method involves the formation of complexes between DNA and a positively charged molecule such as polylysine, on to which is also linked a ligand that is able to bind to a cell surface receptor so that the whole complex can then be internalised by endocytosis.[11]

Vectors for gene delivery—room for improvement

None of the vectors that are available at present is ideally suited for in vivo gene delivery protocols, and this represents a significant obstacle to the successful application of gene therapy to human disease. For instance, systemic in vivo gene transfer to a large number of disseminated metastatic cancer cells would require a high virus titre. To this end effort is being directed at developing new retroviral packaging cell lines that will produce improved titres.[12] In addition, the currently used vectors which are based on murine retroviruses are sensitive to inactivation by human serum, and so strategies are needed to obviate this problem for in vivo clinical applications, for example, by modifying the envelope glycoproteins that are responsible for activation of complement.[13]

It is also desirable to achieve some degree of accuracy of gene delivery to the tumour cells. Attempts are being made to engineer the retroviral envelope to express antibody fragments or ligands that would be able to bind to receptors on specific cell types, thus targeting gene delivery to those cells.[14] For example, retroviral vectors that display heregulin on the envelope may be targeted to breast cancer cells which over-express the receptor HER-2/ *erb-b2*.[15] In a similar way, other viral and non-viral vectors may be modified to target specific cells.[16] Another strategy is to target expression at the transcriptional level by driving expression of the therapeutic gene with a tissue or tumour specific promoter. An example of such an approach is the use of the tissue specific tyrosinase promoter to restrict expression of the transgene to melanocytes and melanoma cells,[17] whereas the use of the promoter of the oncogene *erb-b2* would result in tumour specific expression.[18] Despite the stable integration of retroviral vectors into the host cell genome, expression of the transgene may become downregulated with time. It may be possible to avoid this shutdown by incorporating eukaryotic promoters, such as those described above, in the retroviral vectors.[19]

All the recombinant viral vectors that are being used are replication defective. However, the use of replication *competent* vectors, such as replication competent adenoviral vectors, has been proposed although there would be important safety issues to consider.[20] These vectors have the potential to boost the efficacy of in vivo gene therapy significantly by amplifying the number of cells to which the therapeutic gene can be delivered in vivo.

Gene therapy for disorders caused by defects in a single gene

Adenosine deaminase deficiency, familial hypercholesterolaemia, and cystic fibrosis

These diseases are caused by a single gene defect and will be discussed to illustrate the principles by which gene therapy can be applied. Progress has been made on experimental models of these disorders and encouraging data from these have led to the start of clinical trials. These diseases usually result in death in childhood or young adulthood, and are suitable for the application of experimental treatment approaches because conventional treatment is limited and not curative.

Lack of the enzyme adenosine deaminase (ADA) causes poor function of T and B cells, resulting in a form of severe combined immunodeficiency. The gene encoding ADA can be transferred to autologous T cells ex vivo followed by re-infusion of the gene modified cells. In animal models, the re-infused T cells are found to retain function. Two patients, who are claimed to be the first "successful" cases of gene therapy in humans, have been treated by using engineered autologous T cells. Although these patients were found to show persistence of the re-infused T cells, improved immune function, and some clinical benefit, there is some uncertainty as to whether the clinical improvement seen in these patients was entirely the result of gene therapy because they were also given concurrent enzyme replacement therapy with polyethyleneglycol linked bovine ADA.[21] Other ongoing trials are attempting gene transfer to haemopoietic stem cells in the expectation that this would result in a more prolonged clinical effect.

Familial hypercholesterolaemia (FH) is caused by a mutation in the gene encoding the low density lipoprotein (LDL) receptor, and in the homozygous form of the disease, severe coronary artery disease leads to death at a young age. In animal models, a functioning LDL gene can be transferred to autologous liver cells ex vivo. Re-infusion of the gene modified cells results in stable engraftment in the liver and a decrease in serum lipids is seen. This approach has been extended to humans using hepatocytes cultured from a partial hepatectomy.[22] Stable engraftment of modified hepatocytes was observed, but there was only little or no effect on serum lipids. For there to be useful clinical application, gene transfer to a larger number of hepatocytes would be required. In addition, if an efficient method of delivering the gene to liver cells in vivo were available, the treatment procedure would be simpler. Numerous other metabolic disorders caused by a single genetic defect could potentially be corrected by a similar strategy.

A mutation in the cystic fibrosis transmembrane conductance regulator (*CFTR*) gene, which codes for a chloride channel in the cell membrane, is the cause of cystic fibrosis. Homozygotes have a multiorgan disease, but the major cause of early death is the result of lung pathology where abnormalities in mucociliary clearance lead to recurrent and chronic infection followed by destruction of lung tissue. Hence attempts have been made to deliver the normal *CFTR* gene to respiratory epithelial cells, where it is envisaged that attaining a small proportion of cells with functioning *CFTR* may be sufficient to show clinical benefit. Clearly, efforts at corrective therapy would be most useful in the early stages of the disease before irreversible structural tissue changes occur. Current trials employ cationic liposomes or adenoviruses as vectors and these may be administered as aerosols or by lavage.[23] Although adenoviruses are expected to be more efficient than liposomes, they have resulted in unacceptable inflammatory reactions in some cases. The clinical application of gene therapy for cystic fibrosis is still at a very early stage. As a surrogate functional measure for the transfer of the *CFTR* gene, early trials have monitored variations in potential difference, which reflect changes in ion transport, between the nasal mucosal surface and subcutaneous tissue. The results have been mixed and, although the detection of function in some cases is encouraging, for there to be any useful clinical benefit a long lasting restoration of function in the lower respiratory tract would need to be achieved.

Gene therapy of infectious diseases

Gene therapy for infectious diseases is attractive because the invading organism introduces *pathogen specific* genes which are an ideal target for genetic intervention. For example, antisense oligonucleotides can be synthesised with high specificity for gene targets upon which replication of the pathogen is dependent, but which should not recognise any cellular genetic material.[24] Such approaches have been suggested for treating protozoan parasite infections for which drug therapy is currently inadequate.[24]

Viral infections offer similar opportunities for specific genetic interventions. Thus, in cancers with a known viral aetiology, in the infected cells there will be viral genes that are unrelated to any cellular genes. Therefore, gene therapy targeted against papilloma virus transforming proteins E6 and E7 might be effective in treatment of cervical cancer; similarly, hepatitis B virus (hepatocellular carcinoma), human T cell lymphotropic virus types 1 and 2 (adult T cell lymphoma/leukaemia), and Epstein–Barr virus (nasopharyngeal carcinoma and Burkitt's lymphoma) all offer viral specific targets for gene therapy.[25]

Gene therapy has been also proposed for the treatment of AIDS in the absence of an effective vaccine or drug treatment against the human immunodeficiency virus (HIV).[26] HIV gene expression is directed by a series of regulatory proteins which control levels of viral protein production and the switch from latency to productive infection.[27] One of these proteins, TAT, is required for viral gene expression from the viral promoter in the long terminal repeat. Hence, it may be possible to use the complexity of the control of genome expression against the virus to protect the target CD4+ T cells. For instance, T cells removed ex vivo can be transduced with constructs which use the HIV LTR to direct expression of a suicide gene such as the *HSVtk* gene (see below) (fig 20.4).[28] When these T cells are returned in vivo the absence of TAT will prevent expression of the *tk* gene. If they become infected with HIV, however, the wild type virus will provide TAT in *trans* and expression of the transgene *HSVtk* will be activated. Treatment of the patient with ganciclovir would kill the infected T cells before they could serve as a reservoir of viral production, thereby limiting the ability of HIV to infect more cells. However, such approaches would be unlikely to abolish infection and would, at best, only slow the progression of disease. Other gene therapy approaches have also been proposed which seek to interfere specifically with viral replication steps without killing the infected T cells,[29] including the transduction of CD4+ T cells with TAT dependent, HIV specific, antisense, or ribozyme constructs.[30 31] So far, in vitro experiments have shown promising results in that these constructs can protect tissue culture cells from infection with HIV and applications are currently being approved for trials in HIV infected patients.

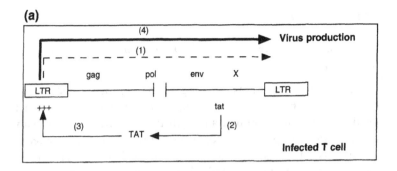

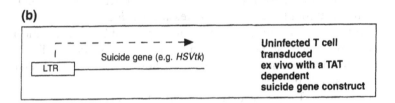

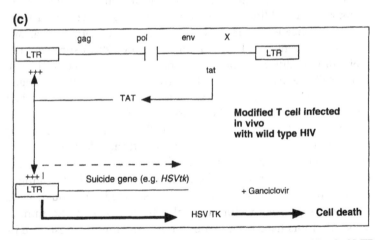

FIG 20.4—*(a) T cell infected by HIV. (b) Uninfected T cell—basal level of LTR transcription is insufficient to produce significant levels of* HSVtk. *(c) Infected T cell—production of TAT by the infecting HIV upregulates transcription of TAT dependent* LTR–HSVtk *construct and the T cell becomes sensitive to ganciclovir. Therefore, infected T cells can be killed in vivo before they produce more infectious HIV.*

Gene therapy of cancer

Immunotherapy: enhancing the anti-tumour immune response using gene transfer techniques

There are observations to suggest that human tumours are immunogenic, because immunosuppressed patients show a higher incidence of some tumours, although these do not include many of the common epithelial neoplasms. Furthermore, spontaneous regressions of metastatic disease from melanomas and renal carcinomas are well recognised, albeit on rare occasions. In recent years, studies using genetic cloning techniques or biochemical elution methods have provided increasing evidence for the existence of tumour specific antigens which are expressed by human cancer cells and which can be recognised by specific cytotoxic T cells.[32][33] An effective rejection response is not, however, mounted despite the existence of these T cells.

Cytotoxic T cells become activated when specific receptors on the T cell recognise a tumour antigen presented with a major histocompatibility complex (MHC) molecule on the surface of the tumour cell. In addition, for activation to occur, other stimulatory signals are needed which are provided in the form of various cytokines secreted by T helper cells. The cytokines stimulate the proliferation, differentiation, and maturation of T cells. The T helper cells themselves need to be activated by professional antigen presenting cells (APC), such as dendritic cells, which have taken up antigens derived from the tumour cells (fig 20.5).

Various mechanisms allow tumour cells to escape elimination by immune cells.[34] Tumour cells grow under strong selective pressures in vivo and therefore tumour cells that lose expression of specific tumour antigens, or any of the molecules involved in the antigen presentation machinery, would be selected for outgrowth. Immunosuppressive factors may also be released by the tumour cells. Attempts have been made to boost the immune response by using recombinant cytokines, such as interleukin-2, but, although there have been successes in a small minority of cases, the systemic administration of cytokines produces severe side effects.[35] Hence attempts have been made to transfer cytokine genes directly to tumour cells, to enable the tumour cells to secrete the cytokine themselves and thus achieve a high constant cytokine level at the local tumour site (fig 20.6).[36-39] Another approach to attain high cytokine concentrations in the vicinity of tumours is to isolate

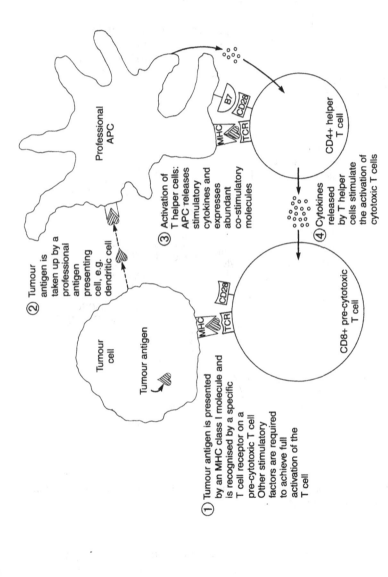

② Tumour antigen is taken up by a professional antigen presenting cell, e.g. dendritic cell

③ Activation of T helper cells: APC releases stimulatory cytokines and expresses abundant co-stimulatory molecules

Professional APC

Tumour cell

Tumour antigen

MHC

B7

TCR

CD2a

CD4+ helper T cell

MHC

TCR

CD2a

CD8+ pre-cytotoxic T cell

① Tumour antigen is presented by an MHC class I molecule and is recognised by a specific T cell receptor on a pre-cytotoxic T cell. Other stimulatory factors are required to achieve full activation of the T cell

④ Cytokines released by T helper cells stimulate the activation of cytotoxic T cells

FIG 20.5—*Activation of an anti-tumour cytotoxic T cell response.*

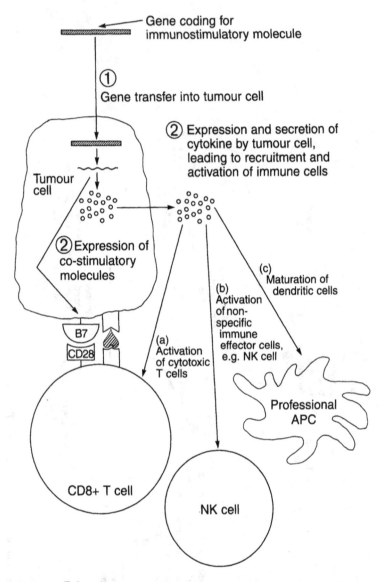

FIG 20.6—*Enhancing an anti-tumour immune response: transfer of a gene encoding an immunostimulatory molecule.*

tumour infiltrating lymphocytes and then to transfer a cytokine gene to these cells ex vivo. When these cytokine secreting lymphocytes are re-infused they aggregate at the sites of tumour cells.[35]

Optimal activation of T cells also requires the interaction between various accessory molecules on the T cell and the antigen presenting cell. Co-stimulatory molecules are expressed abundantly on professional APCs but are not present on most tumour cells. In an attempt to enhance the T cell response and enable tumour cells to activate T cells directly, co-stimulatory molecules such as members of the B7 family, which bind to CD28 on T cells, have been expressed on tumour cells (fig 20.6).[40 41]

An alternative ex vivo strategy involves obtaining a primary culture of tumour cells from a biopsy, followed by cytokine gene transfer in vitro and re-inoculation into the patient. The cells need to be irradiated before re-injection in order to inhibit replication. In this situation the irradiated cytokine releasing tumour cells are, in effect, being used as a vaccine to generate an immune response against the parental tumour (fig 20.6). It is also possible to elicit a T cell response against an antigen by inoculation with DNA coding for the antigen, using either naked plasmid DNA or a recombinant vaccinia virus vector. In this way genes coding for tumour specific antigens may be used as a DNA vaccine to prime an anti-tumour immune response. Such a response may be boosted by co-injection of a gene coding for a cytokine.

A variation on the use of an immunostimulatory gene, and somewhat akin to the use of non-specific adjuvants, is the attempt to express a highly antigenic molecule on tumour cells. For example, direct in vivo transfer of an allogeneic MHC molecule has been shown to result in the inhibition of tumour growth.[42] It is thought that the strong immune response to the antigenic molecule creates an environment that attracts and activates other non-specific, as well as specific, immune effector cells.

A large number of immunostimulatory genes have been used in efforts to manipulate the immune response against tumour cells, and for all of these some anti-tumour effect has been documented in at least some tumour models (see box on page 282). The ability of a particular cytokine to stimulate a local rejection response does not, however, necessarily correlate with its efficacy in generating systemic immunity against the tumour. This efficacy is crucial to the aim of achieving a major impact on the treatment of minimal residual disease by evoking a systemic immune response that would be able to eliminate micrometastatic lesions and prevent death from metastases. It is apparent that the effects of any one cytokine vary considerably among different tumours, and thus in the clinical situation it will not be easy to

Gene therapy of cancer—immunostimulatory molecules which enhance an anti-tumour response

Cytokines
  Interleukin-1
  Interleukin-2
  Interleukin-4
  Interleukin-6
  Interleukin-7
  Interleukin-10
  Interleukin-12
  Tumour necrosis factor-α
  Granulocyte–macrophage colony stimulating factor
  Macrophage colony stimulating factor
  Granulocyte colony stimulating factor
  Interferon-α
  Interferon-γ

Co-stimulatory molecules
  B7-1/CD80
  B7-2/CD86
  ICAM-1

predict which cytokine would be optimal for a particular case. Among the various cytokines, there is currently interest in granulocyte–macrophage colony stimulating factor (GM-CSF) and interleukin-12. GM-CSF has been found to be potent in stimulating systemic immunity to poorly immunogenic tumours, and is thought to play an important role in increasing the number and promoting the maturation of professional antigen presenting cells,[43] whereas interleukin-12 strongly favours the development of cell mediated immunity which is believed to be important for anti-tumour responses.[44]

Prodrug activating enzymes

The gene coding for a viral or bacterial enzyme that catalyses conversion of a non-toxic prodrug to a toxic compound can be introduced into and expressed in a cell, making the cell sensitive to an otherwise innocuous drug.[18 45 46] A widely used system involves herpes simplex virus *thymidine kinase* which phosphorylates ganciclovir, forming an intermediate that is subsequently converted intracellularly to a compound that interferes with DNA replication. Other prodrug/enzyme systems which are potentially more potent are also being developed.[47] An attraction of this approach is the "bystander effect", whereby adjacent cells that do not express the *HSVtk* gene are also killed.[48] In addition, the killing effect may itself promote an immune response against the tumour.[49] These observations suggest that it may be necessary only to deliver the *HSVtk* gene to a small number of cells for there to be an appreciable effect on the tumour.

Tumour suppressor genes and oncogenes

It is possible to introduce a functioning copy of a tumour suppressor gene into a tumour cell or to use antisense technology to disrupt expression of a dominant oncogene.[50] It is far from clear, however, whether such an approach would be effective for the treatment of tumours in vivo. It is unlikely that correcting just one of a number of genes that contribute to the malignant state will cause the cell to revert to a normal state. In addition, it is expected that such an approach will demand that the therapeutic gene be delivered to every single tumour cell in the host, a requirement that current vectors cannot meet. Nevertheless, direct in vivo transfer of the tumour suppressor genes, *p53* or *p21*, has been shown to inhibit growth of tumours in experimental models and, surprisingly, some apparent bystander killing was also seen.[51-53]

Chemoprotective genes

Gene transfer techniques may prove to be useful in cancer therapy in a different scenario: a gene that codes for resistance to chemotherapeutic drugs, such as *MDR-1*, which codes for a drug efflux pump that causes the cell to become resistant to a range of chemotherapeutic agents, may be transferred to autologous haemopoietic stem cells ex vivo.[54] The gene modified cells can be re-introduced to the patient, who will then be able to withstand much higher doses of chemotherapy with less risk of life threatening bone marrow suppression. There is, however, still some debate about how much value increased doses of chemotherapy would be to the patient or if the therapy related toxicity will be transferred to other tissue types.

Gene therapy of cancer—towards clinical applications

Despite the positive results from laboratory models using these various strategies, caution must be applied in directly extrapolating these findings to the situation in human cancers. Transplantable tumours in laboratory animals are much faster growing than most human tumours, which have often been in existence for years or months before clinical presentation. Hence, the immunological interactions between the tumour and host immune cells in human disease may be significantly different from that in experimental models. Indeed, T cells from tumour bearing hosts show signal transduction abnormalities, which may contribute to the generalised impairment of immune function seen in cancer patients.[55] It may be that a combination of the various strategies described above, together with conventional therapies, would prove the most productive way forward.

More than a hundred clinical trials of gene therapy are ongoing worldwide, most of which are for cancer. It must be kept in mind that most of these are early trials where toxicity and transfer of function using surrogate parameters are being investigated. In the meantime, in order to enable gene therapy to make a significant impact on clinical applications, there is an urgent need to improve on vectors for gene transfer.

1 Jolly D. Viral vector systems for gene therapy. *Cancer Gene Ther* 1994;1:51–64.

2 Vile R, Russell SJ. Gene transfer technologies for the gene therapy of cancer. *Gene Ther* 1994;1:88–98.

3 Ali M, Lemoine NR, Ring CJA. The use of DNA viruses as vectors for gene therapy. *Gene Ther* 1994;1:367–84.

4 Miller A. Retroviral vectors. *Curr Top Microbiol Immunol* 1992;158:1–24.

5 Graham FL, Prevec L. Methods for construction of adenovirus vectors. *Mol Biotechnol* 1995;3:207–20.

6 Miller AD. Retrovirus packaging cells. *Hum Gene Ther* 1990;1:5–14.

7 Cornetta K. Safety aspects of gene therapy. *Br J Haematol* 1992;80:421–6.

8 Ledley FD. Nonviral gene therapy: the promise of genes as pharmaceutical products. *Hum Gene Ther* 1995;6:1129–44.

9 Gao X, Huang L. Cationic liposome-mediated gene transfer. *Gene Ther* 1995; 2:710–22.

10 Zhu N, Liggitt D, Liu Y, Debs R. Systemic gene expression after intravenous DNA delivery into adult mice. *Science* 1993;261:209–11.

11 Michael SI, Curiel DT. Strategies to achieve targeted gene delivery via the receptor-mediated endocytosis pathway. *Gene Ther* 1994;1:223–32.

12 Cosset F-L, Takeuchi Y, Battini J-L, Weiss RA, Collins MKL. High-titre packaging cells producing recombinant retroviruses resistant to human serum. *J Virol* 1995;69:7430–6.

13 Takeuchi Y, Porter CD, Strahan KM, Preece AF, Gustafsson K, Cosset FL, *et al.* Sensitization of cells and retroviruses to human serum by (a1-3) galactosyltransferase. *Nature* 1996;379:85–8.

14 Russell SJ, Hawkins RE, Winter G. Retroviral vectors displaying functional antibody fragments. *Nucleic Acids Res* 1993;**21**:1081–5.

15 Han X, Kasahara N, Kan YW. Ligand-directed retroviral targeting of human breast cancer cells. *Proc Natl Acad Sci USA* 1995;**92**:9747–51.

16 Miller N, Vile R. Targeted vectors for gene therapy. *FASEB J* 1995;**9**:190–9.

17 Vile RG, Hart IR. In vitro and in vivo targeting of gene expression to melanoma cells. *Cancer Res* 1993;**53**:962–7.

18 Harris JD, Gutierrez AA, Hurst HC, Sikora K, Lemoine NR. Gene therapy for cancer using tumour-specific prodrug activation. *Gene Ther* 1994;**1**:170–5.

19 Vile RG, Diaz RM, Miller N, Mitchell S, Tuszyanski A, Russell SJ. Tissue-specific gene expression from Mo-MLV retroviral vectors with hybrid LTRs containing the murine tyrosinase enhancer/promoter. *Virology* 1995;**214**:307–13.

20 Russell SJ. Replicating vectors for gene therapy of cancer: risks, limitations and prospects. *Eur J Cancer* 1994;**30A**:1165–71.

21 Blaese RM, Culver KW, Miller AD, Carter CS, Fleisher T, Clerici M, *et al*. T lymphocyte-directed gene therapy for ADA⁻ SCID: initial trial results after 4 years. *Science* 1995;**270**:475–80.

22 Grossman M, Rader DJ, Muller DWM, Kolansky DM, Kozarsky K, Clark BJI, *et al*. A pilot study of ex vivo gene therapy for homozygous familial hypercholesterolaemia. *Nature Med* 1995;**1**:1148–54.

23 Alton EWFW, Geddes DM. Gene therapy for cystic fibrosis: a clinical perspective. *Gene Ther* 1995;**2**:88–95.

24 Miller N, Vile RG. Gene transfer and antisense nucleic acid techniques. *Parasitol Today* 1994;**10**:92–7.

25 Schulz TF, Vile RG. Viruses in human cancer. In: Vile RG, ed. *Introduction to the molecular genetics of cancer*. Chichester: John Wiley & Sons, 1992:137–76.

26 Gilboa E, Smith C. Gene therapy for infectious diseases: the AIDS model. *Trends Genet* 1994;**10**:109–14.

27 Subbramanian RA, Cohen EA. Molecular biology of the human immunodeficiency virus accessory proteins. *J Virol* 1994;**68**:6831–5.

28 Brady HJM, Miles CG, Pennington DJ, Dzierzak EA. Specific ablation of human immunodeficiency virus Tat-expressing cells by conditionally toxic retroviruses. *Proc Natl Acad Sci USA* 1994;**91**:365–9.

29 Dropulic B, Jeang KT. Gene therapy for human immunodeficiency virus infection: genetic antiviral strategies and targets for intervention. *Hum Gene Ther* 1994;**5**:927–39.

30 Altman S. RNA enzyme-directed gene therapy. *Proc Natl Acad Sci USA* 1993;**90**:10898–900.

31 Buchschacher GL, Panganiban AT. Human immunodeficiency virus vectors for inducible expression of foreign genes. *J Virol* 1992;**66**:2731–9.

32 Boon T, Cerottini J-C, Van den Eynde B, van der Bruggen P, Van Pel A. Tumor antigens recognised by lymphocytes. *Annu Rev Immunol* 1994;**12**:337–65.

33 Van den Eynde B, Brichard VG. New tumor antigens recognized by T cells. *Curr Opin Immunol* 1995;**7**:674–81.

34 Vile RG, Chong H, Dorudi S. The immunosurveillance of cancer: specific and non-specific mechanisms. In: Dalgleish AC, Browning MJ, eds. *Tumor immunology*. New York: Cambridge University Press, 1996:7–38.

35 Rosenberg SA. The immunotherapy and gene therapy of cancer. *Clin Oncol* 1992;**10**:180–90.

36 Pardoll D. New strategies for enhancing the immunogenicity of tumors. *Curr Opin Immunol* 1993;**5**:719–25.

37 Blankenstein T. Increasing tumour immunogenicity by genetic modification. *Eur J Cancer* 1994;**30A**:1182–7.

38 Colombo MP, Forni G. Cytokine gene transfer in tumor inhibition and tumor therapy: where are we now? *Immunol Today* 1994;**15**:48–51.

39 Tepper RI, Mule JJ. Experimental and clinical studies of cytokine gene-modified tumor cells. *Hum Gene Ther* 1994;5:153–64.

40 Hellstrom KE, Hellstrom I, Chen L. Can co-stimulated tumor immunity be therapeutically efficacious? *Immunol Rev* 1995;145:123–45.

41 Allison JP, Hurwitz AA, Leach DR. Manipulation of costimulatory signals to enhance antitumor T-cell responses. *Curr Opin Immunol* 1995;7:682–6.

42 Plautz GE, Yang Z-Y, Wu B-Y, Gao X, Huang L, Nabel GJ. Immunotherapy of malignancy by in vivo gene transfer into tumors. *Proc Natl Acad Science USA* 1993;90:4645–9.

43 Dranoff G, Jaffee E, Lazenby A, Golumbek P, Levitsky H, Brose K, *et al.* Vaccination with irradiated tumor cells engineered to secrete murine granulocyte-macrophage colony-stimulating factor stimulates potent, specific, and long-lasting anti-tumor immunity. *Proc Natl Acad Sci USA* 1993;90:3539–43.

44 Tahara H, Lotze MT. Antitumor effects of interleukin-12 (IL-12): applications for the immunotherapy and gene therapy of cancer. *Gene Ther* 1995;2:96–106.

45 Culver KW, Ram Z, Wallbridge S, Ishii H, Oldfield EH, Blaese RM. In vivo gene transfer with retroviral vector-producer cells for treatment of experimental brain tumors. *Science* 1992;256:1550–2.

46 Vile RG, Hart IR. Use of tissue-specific expression of the Herpes simplex thymidine kinase gene to inhibit growth of established murine melanomas following direct intratumoral injection of DNA. *Cancer Res* 1993;53:3860–4.

47 Connors TA. The choice of prodrugs for gene directed enzyme prodrug therapy of cancer. *Gene Ther* 1995;2:702–9.

48 Freeman SM, Abboud CN, Whartenby KA, Packman CH, Koeplin DS, Moolten FL, *et al.* The "bystander effect": tumor regression when a fraction of the tumor mass is genetically modified. *Cancer Res* 1993;53:5274–83.

49 Vile RG, Nelson JA, Castleden S, Chong H, Hart IR. Systemic gene therapy of murine melanoma using tissue specific expression of the *HSVtk* gene involves an immune component. *Cancer Res* 1994;54:6228–34.

50 Mercola D, Cohen JS. Antisense approaches to cancer gene therapy. *Cancer Gene Ther* 1995;2:47–59.

51 Yang Z-Y, Perkins ND, Ohno T, Nabel EG, Nabel GJ. The p21 cyclin-dependent kinase inhibitor suppresses tumorigenicity in vivo. *Nature Med* 1995; 1:1052–6.

52 Leeson-Wood LA, Kim WH, Kleinman HK, Weintraub BD, Mixson AJ. Systemic gene therapy with p53 reduces growth and metastases of a malignant human breast cancer in nude mice. *Hum Gene Ther* 1995;6:395–405.

53 Fujiwara T, Cai DW, Georges RN, Mukhopadhyay T, Grimm EA, Roth JA. Therapeutic effect of a retroviral wild-type p53 expression vector in an orthotopic lung cancer model. *J Natl Cancer Inst* 1994;86:1458–62.

54 Sorrentino BP, Brandt SJ, Bodine D, Gottesman M, Pastan I, Cline A, *et al.* Selection of drug-resistant bone marrow cells in vivo after retroviral transfer of human *MDR1. Science* 1992;257:99–103.

55 Zier K, Gansbacher B, Salvadori S. Preventing abnormalities in signal transduction of T cells in cancer: the promise of cytokine gene therapy. *Immunol Today* 1996;17:39–45.

Index

Page numbers printed in **bold** type refer to figures; those in *italic* to tables or boxed material